Nature's Top 40 – Britain's Best Wildlife

nature's
TOP
40

BBC

BRITAIN'S BEST
WILDLIFE

The Top 40 Sights to See

Mike Dilger

Foreword by
Chris Packham

Collins

Acknowledgements

At HarperCollins I would like to thank Myles Archibald, Julia Koppitz, Kirstie Addis and their design team for making my first complete book such an enjoyable, fun and straightforward process.

Nature's Top 40 is a bizarre and wonderful collaboration of three regional BBC departments from Southampton, Leeds and Norwich, and of a superb production and post-production team of Jane French, Jonathan Bigwood, Paul Greenan, Jenny Craddock, Paul Baker, Jon 'vortex' Valters, Christine Hamill, Catherine Shawyer, Isabelle Hutchins, Mike Lane, Steve Holdsworth, Ron Southern, David Wernham, Terry Wooler, Sue Wilkinson, Simon Marks, Richard Taylor-Jones and Gary Moore. Thanks also to Carla-Maria Lawson and Craig Henderson for commissioning and overseeing the series respectively and to Rob Collis for help with tracking down research books and articles.

Out in the field I would also like to thank Chris Packham, Janet Sumner, Iolo Williams and Sanjida O'Connell for their terrific performances, and the camera crews of Trevor Adamson, Phil Putnam-Spencer, Sean Twamley, Joe Cooper, Martin Giles, Graham Hatherley, Steve Phillips, Manuel Hinge, John McIntyre, Mike Riley, Colin Lang, George Mitchell, Nigel Chatters, Dafydd Baines, Colin Bowes, Robbie Burns, Mark Killingback, Stan Raba, Chris Sharman, Matthew Parker, Mark Dodd and Andy Fraser for making us *and* the wildlife look so darn good!

Lastly, and by no means least, thanks to my partner Christina for being so patient, and to my mum Renee, brothers Paul and Andy and their partners, families and friends for being so supportive and encouraging.

The following abbreviations are used throughout the book:
NNR – National Nature Reserves
RSPB – The Royal Society for the Protection of Birds
WWT – Wildfowl and Wetlands Trust

Dedicated to my dad ... you would have been chuffed to bits

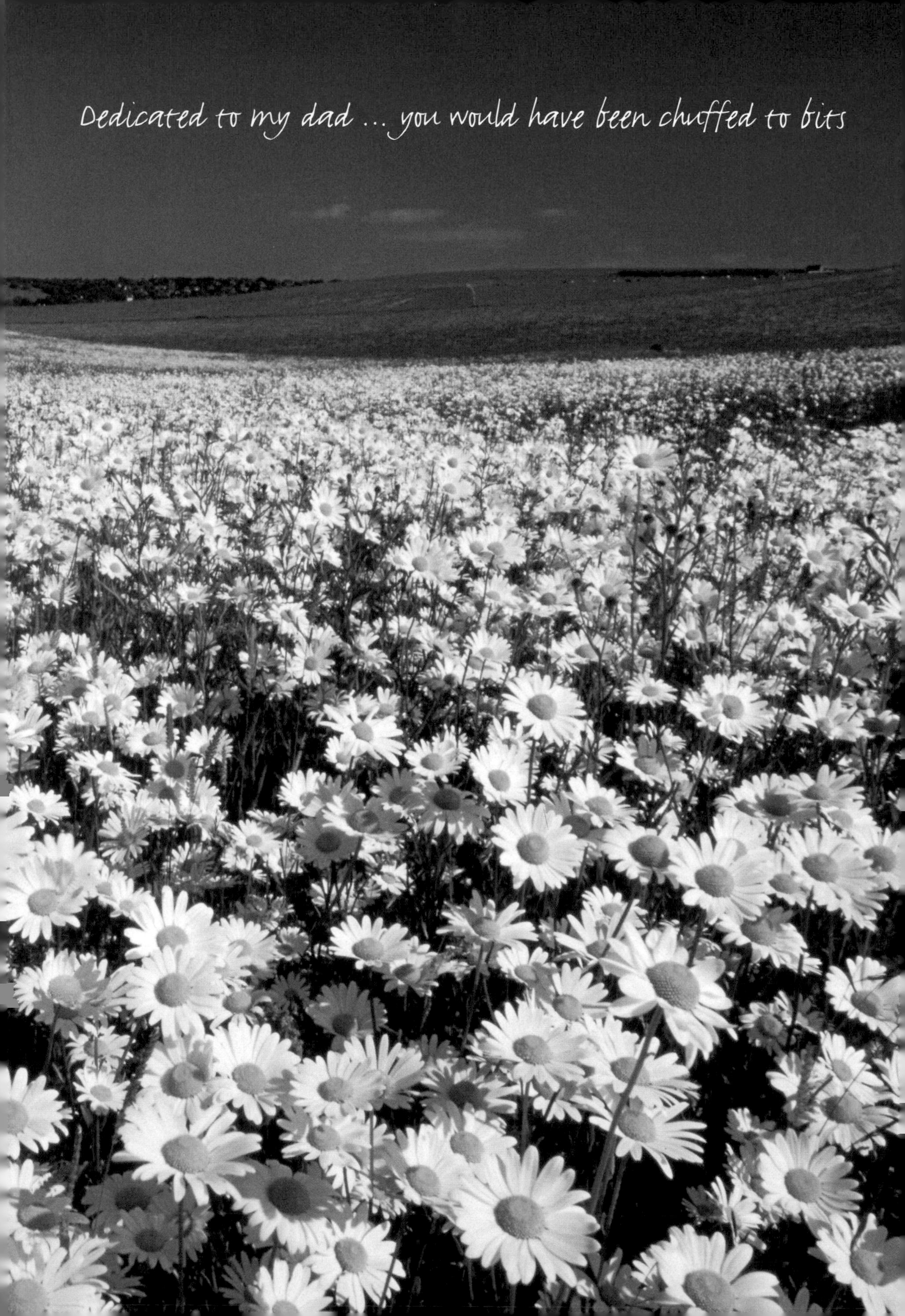

Contents

Foreword
by Chris Packham

I made my first list when I was about eight. I thought nothing of it; it seemed perfectly natural. I was obsessed with reptiles at the time, which seemed normal too, and I'd come down from the heady heights of the dinosaurs to the British Six, as my list was called. Adder, grass snake, smooth snake, sand lizard, common or viviparous lizard and slow worm were the words so neatly inscribed in my book, even if I didn't know what viviparous meant and had to be careful spelling it. I had four out of six ticks alongside them, not bad thinking about it now, but seriously disappointing for the obsessive young naturalist who had to wait an eternity – until he was twelve – to put the final mark alongside sand lizard. Lying in bed that May night wrapped in the warmth of smug satisfaction, I then wondered if I could be the youngest person ever to see the all of the UK's native reptiles.

Several things are revealed by this admission: that list-making is a very important part of any naturalist's behaviour; that an incomplete list can cause real distress and lead to an increase in obsessive behaviour; and that lists have a strong competitive component. When you think about it, they are only made to be completed. Shopping lists, wedding lists, they are all made up of things that we need, and the difference between want and need is a notable one. So, listing things is sometimes helpful but patently questionable, and I'm not sure that it makes you particularly happy sometimes, either. But I am certain that it gets you out of bed and that it can act as a wonderful fuel to get your life lived more fully. Oh, and people say that listing is geeky. So what? If you've picked up and paid for this book you are at least a little bit geeky, and that's good; be proud of it!

It's tempting to imagine that naturalists invented listing, but I'm sure it would be hotly contested from many sides. Historians are also consummate listers, and collectors of anything are listers par excellence, but I think we hold our own. I have a British bird list, a world bird list, a garden bird list, a dog-walking bird list, a garden moth list, a British butterfly and bat and amphibian list, an orchid list, a travel list … and, to be fair, that's nothing. I'm an amateur when it comes to listing. But these are all empirical lists; there are also the subjective lists – and these are very exciting because they are dynamic and more interactive. We brag about the former, but we debate the latter.

My Top Ten Favourite Films Of All Time (the last bit is fabulously childish, but necessarily demonstrative, and should follow all such list titles) obviously bears no resemblance to the list from which I was trying so hard to tick smooth snakes and sand lizards, although Raquel Welch of the *One Million Years BC* era still features in several lists. And it can be thrown into turmoil by a single release; it can require a systematic reappraisal, taking hours and contentious 'discussion' with my movie guru James. And that is great, too, because opinionated lists must be argued about as a matter of honour. I mean, how can this self-appointed 'guru' have *Blade II* in his Top Ten? Or, more to the point, how the hell did black grouse lekking make it into the Top Five of this list when glow-worms are languishing in the 30s? And who seriously thought: it's right that a spider's web is not Top Ten material? And, to my mind, puffins should not be seen or heard in any list, gaudy little chavs. Starlings, swirling about going to roost, good, but so passé. That's like still having *The Italian Job* in your Top Ten Movies. I mean, come on, update will you?

What is even more exciting about *Nature's Top 40* than the order is that it's a UK-based list, and that it finally gives us a chance to bunch together a set of spectacles, which undeniably deserve their status, if not their final positions, in your or my opinion! It proves beyond doubt that the UK is not the land of 'little brown jobs', or 'quite good for its seabird colonies', or 'has a few nice spots'; it is rich in things that can stop you in your tracks, make your heart miss a beat, make you hold your breath, make you travel a hundred miles, make you want to shout out loud, make you make lists of things it will make you do!

Joking aside – and I know that many of you will immediately count through to see how many of this 40 you have already seen – the actual 'final' positions don't matter. What does is that you use this list to get you out to experience and enjoy these spectacles first-hand and that you get some youngsters out there with you. Ask yourself this: how many twelve year olds today have seen all the UK's reptiles in the wild, or, more importantly, lie awake wanting to? If this list is to endure, we need some more of these apprentice listers, and that could be down to you.

Chris Packham
2008

PS The answer is ... I've seen 37. I haven't seen the roosting parakeets or the wild goats, and I would most like to get to grips with adders dancing.

Introduction

Music charts lend themselves to a list according to their popularity based on the solid statistics of sales volume, as do films, and of course computer games and books ... but wildlife? Some may wonder how we dare make a Top 40 list of Britain's greatest wildlife spectacles in such a manner, and may feel that the act of giving each of our chosen entries a number might even cheapen the very spectacle we have filmed and written about.

Others will be indignant that their favourite spectacle has unfairly been demoted to the lower regions of the Top 40. But look down any of the numerous Top 40 lists that have formed the basis of a variety of television programmes, such as the *Funniest Moments on TV*. Undoubtedly part of the fun is exclaiming in a faux-indignant way that the clip of talk-show host Russell Harty being attacked by Grace Jones is not as funny as the lower-ranked, but hilarious, moment where a slightly-worse for-wear Delia Smith screams, 'Lets be having you!' to a bemused Norwich City football crowd at half-time. There will also be those wildlife purists not best pleased that immigrant (and therefore 'lesser') spectacles, such as 'parakeet roosts' and 'rutting goats', have audaciously elbowed a spot in the list and now vie for attention alongside our native British spectacles. And, come to think of it, why are badgers playing and kingfishers fishing completely absent from the list altogether?

Our Top 40 was compiled from contributions by members of the public, following a request on the BBC *Nature's Calendar* website for their suggestions. The 40 most popular suggestions put forward were then 'moulded' into an order by a panel of wildlife experts who argued (I believe well into the night) on the relative merits of each species and exactly why, for example, the thrilling clouds of butterflies, which was positioned at no. 27, deserved a higher spot than the enchanting light show put on by glow-worms at a more lowly number 38.

The factors taken into account when compiling this list included a combination of how unique the spectacle is to Britain, and a 'thrill-ability factor'. Some of the entries in our 'wildlife chart' involve huge numbers of one species, such as pink-footed geese returning to roost, or bluebells flowering synchronously in a spring wood; others comprise either fewer or single individuals with particularly remarkable or fascinating behaviour, such as spiders building webs or great crested grebes courting. The best spectacles inevitably involved large numbers of one species (or a combination of species) acting in a remarkable manner, with Britain additionally being the best place in the world to view that event. The prime example of this is gannets diving,

which deservedly made our number 1, because Britain holds an astonishing 63 per cent of the world population of gannets, and the very sight of flocks of these birds pelting into the water is frankly breathtaking.

It is important to bear in mind that these spectacles have not, of course, evolved for our viewing pleasure, and we are nothing more than voyeurs in what serve as vital functions in the mating and survival games of each of our entries. So, in addition to helping you find out more about how to come across each spectacle, the body of the book is primarily written to enable you to understand exactly what is going on and why, which should hopefully enhance your enjoyment and renew your appreciation of the wonderful wildlife still eminently viewable on and around the British Isles.

We make no apologies for the order of our Top 40; you may not agree with it but, hopefully, it may just occasionally form a stimulus for conversation in place of the latest TV series plotline. Perhaps you will be encouraged to make your own 'Wildlife Hit Parade'. The primary motive behind the series and this book is, unashamedly, to encourage people to jump off their sofas, turn off their television sets and stow away the computer games console in order to get some fresh air in their lungs and a few spectacles under their belts instead.

Finally, with some insider information, the vast majority of these spectacles can be seen with a little planning and the requisite luck. Only when the joy or 'Gospel of Wildlife Watching' spreads to as many people as possible (irrespective of the order in which we place them), will these plants, the animals and their habitats be truly cherished, valued and conserved for future generations.

Natterjack toads

WHEN
April to mid-May

WHERE
Ainsdale NNR, Merseyside;
Caerlaverock Nature Reserve
(WWT), near Dumfries

40 Mating natterjacks

The 'Birkdale nightingale', 'Bootle organ' and 'Thursley thrush' are all regularly used colloquial monikers that, in certain regions, have replaced the more commonly accepted name of our smallest and rarest British toad, the natterjack. The reason why such a seemingly inconspicuous and rare toad should been given so many local nicknames is purely down to its incredible ratchetlike call, which marks the highlight of its breeding season, and is also a tremendous spectacle for anybody with a penchant for toadspotting.

This diminutive relation of our common toad is entirely restricted to Europe, with its heartland being the Iberian Peninsula, and becoming progressively rarer further north. In Britain the natterjack was widely if locally distributed around southern and western coastal locations, but healthy populations can now only be found along the northwest English coast around Merseyside and in southwest Scotland centred on the Solway Firth.

To the untrained eye, natterjacks are similar to the common toad, but their size of no more than 75 millimetres, dry brownish to olive-green warty skin and yellow stripe, which runs the length of their backbone like a 'no-parking here' line, easily distinguishes this wonderful and mercurial little toad. In addition, natterjacks have short hindlimbs, giving them the ability to run at surprising speeds over short distances. Unlike common toads, where the females are generally larger, there is little difference in the size of the natterjack sexes, even when the females are bursting with eggs and ready to spawn.

Natterjacks in Britain are now found almost exclusively among sand dunes and the periphery of salt marsh close to our coastlines; they will always be below 100 metres above sea level. Befitting a species that is most abundant in Spain, it is no coincidence that this is one of the warmest habitats in Britain. These sandy spots are also perfect for a species that is a compulsive burrower, meaning that natterjack toads are easily able to dig down to escape from the extremes of temperature. The short, dense vegetation attracts lots of invertebrate prey, and, as sandy places are also well drained, natterjacks have evolved to breed in ephemeral freshwater pools and ponds in the dune slacks. A typical breeding site will often be no more than a small, sandy, shallow and unvegetated pool with a maximum depth of 30 to 50 centimetres that will often have completely dried out by the height of summer.

In common with all our native amphibians and reptiles, the natterjack is a species that opts out of the coldest autumn and winter months by hibernating. In the case of natterjacks, this time is spent underground in self-excavated burrows either alone or with other toads, with the largest number – an astonishing 44 – recorded by the celebrated herpetologist Trevor Beebee.

'Do not park here! The solid yellow line is the key identification feature of the natterjack toad.

On good nights you might get the impression that you had dropped in at the Okavango delta or the Amazon basin at dusk rather than it being just a misty night on Merseyside!

After this period of torpor, the toads then emerge into the light in March or early April once the air temperatures have warmed up sufficiently to sunbathe.

Adults leaving their hibernacula, or winter residence, have usually only two things on their mind, and the first of these is food. The toads emerge to start foraging at dusk and may move several hundred metres from their burrow to feed in the dune slacks before returning to the same burrow before dawn. The natterjack is much more athletic than the common toad and hunts actively by running down its prey over short distances. The long sticky tongue then shoots out at lightning speed to ensnare the unfortunate ant, bug, beetle or fly. Smaller prey is swallowed immediately while larger prey often takes a while to gulp down and can be disposed of by the toad's ingenious ability to retract its eyeballs back into its head, thereby applying pressure to the roof of the mouth and, hence, helping to crush and swallow the food.

While natterjacks will eat virtually anything that will fit into their mouths, the adults themselves have comparatively few predators due to the nature of their skin, as the larger warts contain parotid glands that secrete a poison when molested. This deters most predators, with the exception of some members of the crow family, which have learnt to disembowel them leaving the skin behind, and grass snakes, which seem less susceptible to the poisons. A secondary defence for the natterjack when faced with a grass snake is to puff itself out like a little balloon and stand high on all four legs to give the impression that it is larger and more menacing than it really is.

Having fed, the toads are then keen to move on to the details of mating. The first stage involves the short migration back from their winter quarters to the breeding site of choice and this generally occurs in late March or early April on evenings where the temperature is above 8°C. The males are the first to arrive at the pond and initially tend to occupy burrows close to the water. It is easy to tell when the breeding activity starts in earnest as the natterjack has the distinction of being Europe's loudest amphibian and its calls can easily be heard from a mile away on a warm, still night. After sunset the males emerge from their burrows to take up evenly spaced positions in the pond margins. They then adopt a stance with their forelimbs straightened to keep their head and, more importantly, their vocal sac clear of the water.

The sight and sound of a male in full voice is an unforgettable experience. Its strident call is produced by an inhalation of air through a couple of slits in the bottom of the mouth, which is then shuttled backwards and forwards between the lungs and the vocal sac. This results in an enormous inflation of the vocal sac so the toad's actual head becomes three times its normal size. As the normally purplish or bluish throat distends, in the light of a torch it appears ghostly white and translucent and is very reminiscent of a child's bubblegum bubble. Normally, a male will be stimulated into calling by the churring of other surrounding males but even passing vehicles and planes can initiate a chorus. The call sounds like a 'rrrrRIP' that lasts for about a second before a slight upturn at the end, and individual males will usually call continuously for around a minute before taking a short rest and winding up again.

Calling males vastly outnumber the females on any given night, as the males may well stay for the entire breeding season, and a visitation by the female could be confined to just a few hours on one night of the year. As the ratio is skewed towards amorous males, much time is often wasted in chasing, grabbing and releasing other males in a case of mistaken identity. This error is soon sorted out by the grabbed male, who makes a small croak to let the other male know it is 'riding the wrong chariot'; he is then immediately released. Females do not often arrive until after dark and are usually grabbed by the first males who chance upon them. In each case the male then proceeds to lie on the back of the female and keep hold by using his forelimbs to tightly clasp around her armpits in a grip called 'amplexus', which is strengthened by rough nuptial pads on the digits of the toad's forelimbs.

This coupling process is a lot more gentlemanly than with common toads, where an unmated female can become covered by a writhing mass of males to the extent that she may even occasionally drown. With natterjacks, however, there seems little territorial behaviour or aggression, and, once a male has attached himself to a female, it seems often to be taken as read that the female is no longer available and the couple are left unmolested as the female selects a suitable shallow spot to begin spawning.

Most spawning takes place at night but can occasionally be seen in broad daylight after particularly busy nights. The egg-laying itself is a protracted process taking around three hours, with the female ejecting eggs in a long string, in between periods of rest; the male then fertilises these externally. The natterjack females will produce between 3,000 and 4,000 eggs in this way; the strings are easily distinguished from those of the common toad after a period of 24 hours, as the eggs develop into a single row as opposed to the double row of their common counterpart. As soon as the spawning is finished, the male swims away to look for more mating opportunities, while the female leaves the water, spent, her breeding season over and wanders off to feed.

During sunlight hours, natterjack tadpoles are visible in large numbers at the water's surface or along the pond edges. This contrasts with frog tadpoles, which are much more shy and retiring. The tadpoles must grow quickly as they will all die if the pond dries out completely. Those that survive the high summer droughts and the jaws of predatory insects – the minority – metamorphosise, and head to the land and a brave new world in which they must fatten up in preparation for hibernation.

39 Wood ant nests

Ants are a subject to which many people have never given much thought, unless it is how to extinguish them when they invade our houses. The best thing to do, though, is to take a moment to watch ants, rather than exterminate them. There is much to be admired about an organism with levels of societal organisation that are the envy of the natural world.

Britain's ant fauna of around 50 species is impoverished when compared to the tropics; the famous biologist and ant expert EO Wilson once stated that he could find as many species of ant on one tree in the Peruvian Amazon as exist in the whole of the British Isles. However, what Britain lacks in quantity, it makes up for in quality, with the wood ants being among the most charismatic of this immediately recognisable, but little understood, group of insects.

Of the six possible species of wood ant in Britain, only three are commonly encountered. The Southern wood ant, the Northern hairy wood ant and the Scottish wood ant all look quite similar and are identified by the shape and amount of hair on their heads or, rather more easily, by geographical location, with the exception of a few places where two of these species will occasionally overlap. Moreover, none of these species will come anywhere near your larder!

With the largest individuals being around ten millimetres in length, wood ants are not only the biggest British ants, but they also have the most populous colonies, with some nests reaching close to half a million individuals. They are also the only true woodland ants in Britain, as all of our other native species need warmer temperatures and so are invariably found in more open and warmer habitats such as heathland or grassland. The ability of the wood ants to conserve heat in the cooler forest environment make them the ideal candidates for living in what might be considered an unsuitable habitat.

A large wood ant nest on an otherwise fairly bare forest floor is an impressive sight. These domes can easily reach to over two metres in height and are, moreover, only the above-ground part of the nest; the structure can extend at least as deep again underground. If one of these nests were to be sliced in half, it would reveal a citadel of complex interconnecting tunnels and galleries that would surpass even the wildest imagination of today's science fiction writers.

During the summer months the surface of these nests can be a wriggling mass of thousands of ants busily carrying out their specific chores. The vast majority will be wingless workers, essentially immature females that take no part in reproduction. These differ in size between five and ten millimetres, according to the different jobs that have been preordained to carry out from the moment they emerge from their pupae. Some will spend their lives collecting honeydew – a saccharine substance found on the leaves of plants – or catching insect prey, while others will tend to the eggs, larvae and pupae, and a still

Wood ants

WHEN
Most active from early spring through to late autumn

WHERE
In scattered woodlands and coniferous plantations throughout Britain; distributions depend on species. Coed y Brenin, near Dolgellay, Gwynedd; Glen Affric NNR (SNH), Inverness

different caste will have the responsibility of building and repairing the nest.

The sole egg-laying machine in the wood ants' nest is the queen. She can usually be identified by her much larger thorax that houses the flight muscles, and an outsized abdomen, which contains the ovaries and a sperm sac from her single mating. Although the vast majority of her eggs will develop into the different worker castes, during spring, winged queens and males begin to emerge from slightly larger pupae. These 'sexuals' will take to the air and mate, after which the queens disperse to form new colonies away from the mother nest. Separate nests in the same wood are able to synchronise the release of their sexuals on the same day so that new populations have a shuffled combination of genes to avoid inbreeding. These synchronised flights of the winged wood ants will only occur during the right climatic conditions and the number produced can be vast as they rise into the canopy to mate on the wing or in the treetops. While many will be eaten, for example by insectivorous birds, the local predator population often becomes quickly swamped by this sudden surplus of food meaning that there is negligible overall impact.

An ant citadel, with a population slightly larger than Bristol and, arguably, far fewer traffic jams!

After this mass aerial ant orgy, the males die and the females return to the ground, shed their now-obsolete wings and look for suitable nest sites. In the case of the Southern wood ant, the queen will track down a colony of a closely related species, the common black ant. She will then gather a few common black ant pupae and construct a special cell within the colony for them; upon hatching they will join their 'stepmother' in killing the original host queen. They will then look after the queen Southern wood ant's eggs but, once the queen's own eggs have hatched, the new Southern workers take over the colony and extinguish the original black ant colony!

The one and only mating from her nuptial flight will have provided the queen with enough sperm to last her entire life, which, in the case of wood ants, may be an astonishing fifteen years, during which time she may well lay hundreds of thousands of eggs. These eggs are placed in the warmest part of the nest until they hatch into hairless larvae. The larvae are then fed on a part-digested liquid meal by the workers, before they finally graduate on to prey items collected from the forest. As the larvae grow, they are meticulously looked after by the specialised workers while they undergo a number of moults until they are ready to pupate into either the workers or the sexuals.

Wood ants also use their incredible social organisation when, after a period of winter hibernation, the first few days of spring will be marked by an increase in activity and the workers can be seen pouring away from the nest along special pathways that they meticulously keep obstacle-free. In contrast to many ant species, which use chemical signals or pheromone trails to find their way, wood ants exploit their excellent eyesight; they use a combination of the shapes of the surrounding vegetation against the sky and the orientation of the sun. Foraging parties will then retrace their route back to the nest with their spoils.

Diligent ant researchers have extrapolated that, on a single warm summer's day, a typical wood ant colony can bring back as many as 60,000 individual items to the nest; these include aphids, caterpillars, beetles, flies, harvestmen, woodlice and plant material. A daily haul of around 140 grams of solid food can have a serious impact on the wood, leading to so-called 'green islands' around nests, where the vast majority of the defoliating insects have been removed, meaning that the leaves remain virtually unblemished. A very important additional food source for wood ants is honeydew which is collected by the ants from aphids that have tapped into a plant's sap. The wood ants will often protect their aphids from attack by ladybirds and, in return, effectively milk them, like we would our cows, for a reward of a tiny drop of sugar solution.

A substantial proportion of the daily items brought back will consist of leaves and twigs, or needles if the nest is under conifers, which will be used for running repairs to the nest. The nest is designed to keep out the rain, but is also constructed to catch heat to enable the ants to operate under conditions much warmer than that of the surrounding ambient air temperature. On the surface of the mound the temperature may vary by as much as 15°C in summer, but the heart of the nest can be kept at an incredibly constant and snug 25°C. Particularly in the first cold but sunny days of spring, it is thought that many ants will emerge to sunbathe before carrying the heat below to warm the nest.

The wood ant workers also have three formidable weapons to defend the nest against potential intruders and to subdue their insect prey. Like all ants they have relatively powerful jaws and huge supplies of formic acid, which can both repel borders and act as an alarm signal to the colony. Finally, the wood ants are able to rely on sheer numbers to help them overrun and subdue the enemy. If the surface of the nest is stirred up, large numbers of ants run to the disturbed point and curl their abdomens around to squirt the acid at the potential threat. Despite these defensive measures, significant numbers of ants will still often be consumed by green woodpeckers and badgers.

Away from the nest, territories may spread for close to 100 metres in all directions depending on the local competition. Ants can recognise colleagues from their own colony by a 'nest odour' and may viciously attack strangers; spring battles sometimes occur between some of the separate wood ant colonies as they delineate territories. Like all civilised societies, diplomacy will usually predominate if there is more than enough territory to go around, resulting in nests of many sizes and stages of maturity in favourable woods.

Glow-worms

WHERE
More common in the south.
Slapton Ley NNR, Devon;
Barnack Hills & Holes
NNR (Natural England),
Cambridgshire; Aston
Rowant Nature Reserve,
near Stokenchurch,
Buckinghamshire

Displaying glow-worms

There can be few more delightful surprises for a naturalist than to walk along a country lane on a summer's evening and to chance upon small pinpricks of luminous light emanating from the fields flanking the hedgerows. In addition to these unique biological lights being one of Britain's most underrated wildlife spectacles, their function plays a vital part in the mating game of a fascinating insect.

The glow-worm is Britain's only bioluminescent representative of an order of insects called fireflies, a group that reaches its maximum diversity in the tropics. However, both their name and the order to which they belong can be very confusing, as glow-worms are neither worms nor flies, but members of the order of beetles.

While those lucky enough to have seen this miniature version of the Blackpool lights are few, even fewer people realise that the light show is merely the final two per cent of the glow-worm's intricate life cycle. For the previous 15 months, they have been quietly transforming from egg, to larva, to pupa and finally to adult as they prepare to go out in a blaze of colour.

In common with the vast majority of insects, the glow-worm starts life in the autumn as an egg. Batches of between 50 and 100 eggs are laid under vegetation or stones by a female from the preceding generation, a task so arduous that the exertion is her last task before she dies.

Each egg takes around 35 days to hatch, after which out pops a larva. The creature that emerges is a lean, mean, killing machine, and, once its skin has hardened, it has only one thing on its mind: food. The glow-worm larva is a specialist eater, restricting itself to a diet of snails and slugs, which are tracked down at night with its sensitive antennae and palps (feelers). Once potential food is located, the larva is able to tackle prey, such as snails, much larger and heavier than itself. It does this by giving the snail's foot a series of nips with its sickle-shaped mandibles (or jaws). These mandibles are hollow and deliver a powerful poison, which firstly acts upon the snail's nervous system to cause paralysis, and secondly starts to digest the snail into a 'soup' that is then lapped up by the larva.

During this process, the snail is paralysed but still alive; very occasionally, partly eaten snails have been known to recover as the toxin wears off and crawl away to fight another day. The larva is capable of eating substantial meals at one time meaning a rapid growth rate. However, this presents a problem as the growth occurs inside an inflexible exoskeleton. The larva solves this like many other immature insects, by undergoing a moult, in which the old skin is shed and replaced with a larger version.

Even though it is the adult females that are associated with the light show,

The flightless female puts on one of nature's finest light shows ... and all to attract a male!

the larva is also able to produce a faint light, which often pulsates on and off. The reasons for this are not entirely clear, but this technique is thought to be used to scare off potential predators. After a couple of moults and as winter takes hold, the larva becomes more lethargic and then hibernates under logs, stones or leaf litter. In the spring of their second year, the larva emerges hungrier than ever and eats voraciously over the spring and summer, undergoing moults along the way in preparation for breeding the following summer. Towards the end of the summer in their second year, as they curl up to see out their second winter before breeding, the much larger females can easily be identified from the males.

By the following spring, the larva is virtually full-grown and may not even need to moult again before it pupates. Once early summer arrives, the larva often loses its nocturnal habits as it actively searches for a suitable place to pupate. This is particularly important for the females: they are remarkably sedentary as adults so use this period to colonise new areas.

The larva then picks a secluded area and begins to pupate, a remarkable process in which the larva's body, including all the organs, is completely broken down and then reconstituted as the adult form takes shape. This process takes just over a week for the females and slightly longer for the males as they have to undergo a more drastic alteration due to their need for the power of flight.

The first glowing females of the season are usually seen from mid-June, and, with their segmented bodies and absence of wings and wing cases, retain more than a passing resemblance to the larvae. The light organ is positioned on the underside of the abdomen and consists of two luminous bands and a couple of luminous spots set either side of the ovipositor, the organ from which the eggs will be deposited. The glow-worm's light is produced by a string of chemical reactions between a small protein called luciferin and a large enzyme called luciferase. Though the complex reaction is still not totally understood, it is an incredibly efficient process: 98 per cent of the energy is released as light, compared to a measly 5 per cent in a light bulb, in which the vast majority of the energy is wasted as heat. This means that the light organs of the glowing females are completely cold to the touch.

The glowing performance usually commences soon after dusk at around 10–10.30 p.m. It is thought the turning on of the light is triggered by a drop in light intensity below a certain threshold level, which explains why glow-worms that advertise in the darkness of a woodland edge will begin glowing earlier than populations in predominantly grassland locations. The female will generally display close to the ground, or up to a maximum height of around

Rough grassland with little light pollution is essential if you are a female glow-worm desperate for attention

It is only virgin females that glow; once a female has mated and begins egg laying, the light organ has served its purpose and is switched off.

40 to 50 centimetres if she feels a higher vantage point would be more beneficial. Females will even glow during rain, but usually stay closer to the ground during inclement weather.

Because the light organs are set on her underside, the female also has to twist her abdomen around to make sure that any males flying past will see her lights. This twisting is often accompanied by a swinging of her abdomen from side to side like a metronome, which, to complete the exhibition, gives the effect from a distance of the light brightening and dimming.

The display usually lasts for a couple of hours, after which, if she hasn't been successful in attracting a mate, she turns off the light and retreats back into the grassy tussocks to prepare for a repeat performance the following night. The females are remarkably sedentary and are often seen displaying from the exactly the same spot until they either snag a mate or die an exhausted spinster after around ten consecutive nights of lighting up the night.

The numbers of glowing females can vary enormously from just a few females at small sites, to Brush Hill in Buckinghamshire, the location thought to contain the largest colony of glow-worms in Britain, with 320 females counted in one visit in 2007.

The male glow-worm looks very different to the female, because he has to be mobile in order to track down the displaying females, and so has a fully functioning pair of wings tucked away under his wing cases. He normally takes flight shortly after the females have started glowing and flies a couple of metres above the ground until he spots a virgin female. He will then drop out of the sky with unerring accuracy next to the female and attempt to climb on her back. It is not uncommon to have a number of suitors chasing a single female, meaning a form of rugby scrum can sometimes ensue as they jostle for position.

Like the adult female, the male cannot eat and so has a very limited adulthood. Once mating is over, the exhausted male dies, usually no longer than a week after emerging from his pupa for his date with destiny. His death is closely followed by the female's after she has dutifully discharged her eggs to produce the next generation.

Although found all over Britain, glow-worm colonies are most abundant in southern England. They can be seen in grassland of every type, apart from sites that have been 'improved' with fertiliser or heavily sprayed with insecticides, and also occur in moorland, heathland and occasionally woodland. The spectacle is mostly a rural phenomenon, and country anecdotes abound of glow-worms being put in jam jars to read by at night!

The last 50 years are thought to have seen a steady fall in the number of colonies thanks to the usual lethal cocktail of habitat destruction, fragmentation and pollution. Artificial lights may well also present a problem as the males could be distracted by the lights and find it difficult to spot the females in the glare. Let's hope, however, that these magnificent insects continue to bring light into our lives for many years to come.

Fox cubs

WHEN
Late April until the end of
June

WHERE
Widespread and can be seen
anywhere, although easiest
seen in cities such as London
and Bristol

37 Playing fox cubs

You would think there could scarcely be enough room for a medium-sized native carnivore to live alongside us in Britain. Yet the cunning and resilient fox has led to it not just surviving, but actually thriving, anywhere from on rural farms to in the heart of Britain's biggest cities. It's a tough life, though, and particularly among the urban residents it's a 'live fast die young' scenario, where cubs must learn the tricks of their trade quickly to give themselves a chance of breeding the following year.

The word 'fox' is considered a very old English word that came from the proto-Indo-European word '*puk*', or Sanskrit '*pucca*', which both mean tail. Our only native canid (member of the dog family) was widespread in Britain from the end of the Ice Age: evidence of fox remains reveals that the earliest human inhabitants hunted them for fur and meat. Despite a history of persecution through the Middle Ages, the number of foxes was scarcely reduced until the rise of pheasant shooting in the Victorian era, when an army of gamekeepers was employed to wipe out the 'vermin'. The liberal use of vastly improved guns, traps and poisons meant that, at the turn of the 20th century, foxes had been virtually exterminated from much of East Anglia and the large estates in eastern Scotland. But, as gamekeeping declined after the First World War, fox numbers recovered, and current estimates indicate the population has remained largely stable over the last 30 years at a pre-breeding population of 250,000 adult foxes.

Despite foxes being recorded from the length and breadth of mainland Britain, their distribution is far from even, with the highest densities occurring in southwest England, the Welsh Borders and up into southern Scotland. While foxes can be found anywhere from moors or woodlands to the centre of towns, they prefer fragmented habitats that are able to provide them with a wide range of cover and plenty of boundary edges along which they can hunt. Contrary to popular belief, despite the relatively recent colonisation of towns by foxes from the 1930s, 86 per cent of foxes are still thought to prefer living in the countryside, although a number may regularly move between the two.

The adult fox and its cubs are immediately identifiable but, on close inspection, many people are surprised by how small foxes actually are. A male dog fox weighs little more than 6.5 kilograms with an average body length of 67 centimetres plus a bushy tail adding a further 40 centimetres, while the female or vixen weighs even less, only marginally more than a domestic cat.

Their coat can vary in both colour and condition during the course of the year and they generally look at their scruffiest in the summer during their long protracted annual moult that begins in April. It is not until autumn that the old fur has fallen out and a new, shorter coat is revealed underneath. By the end of October or early November it is long, thick and ready for the winter.

A fox cub seemingly without a care in the world, but it can be a surprisingly short and brutal life.

In terms of sight, foxes do not enjoy the palette of colours available to the human eye and are often reliant upon movement for the object to register on their visual radar. However, their hearing at low frequencies is particularly acute, and is heavily used at dusk or night-time to track down the rustling of small mammals in the leaf litter. Once the sound is pinpointed, the fox will pounce on an unsuspecting mouse, vole or rat from as far as two or three metres away. A fox's world is also dominated by smells, which are used to track down the next meal. Areas around the fox's territory sprayed with urine are also capable of conveying a range of information about the owner, such as their identity or reproductive state.

While many sightings in both rural and urban areas are of solitary animals, most foxes are part of a group. Most consist of a clear hierarchy with a dominant dog fox and vixen, which will usually be the only pair to breed, a number of mostly subordinate females (female cubs from previous years that have not left the territory) and unrelated males. The number of subordinate foxes within the group will depend both on whether food is plentiful and on the local level of persecution, with favourable conditions leading to groups with as many as ten adults in addition to the alpha pair's cubs.

The groups' territories can vary enormously in size, with rural foxes generally making use of at least 1.5 square miles per group, as opposed to urban foxes where food is more easily acquired, which may have five territories crammed into each half square mile. In upland areas, where fox densities are lower and food is more difficult to locate, the territory may be as large as 12 square miles. The dog foxes will constantly man the borders of their territories after the cubs have dispersed in the autumn and in winter when the females are approaching oestrus. Upon confrontation with the neighbours, who are not deterred by a snarling match, the resident fox will frequently resort to fighting by rearing up on its hind legs and engaging in pushing and biting matches to try to drive the intruder away.

Foxes are able to mate only ten months after birth, with the mating peak occurring early in the New Year when the females come briefly into oestrus. A copulating pair can sometimes become locked together for up to an hour, a feature unique to the dog family; it is a time when both foxes can be left very vulnerable. This mating period is also the time when the bloodcurdling screams of the vixen and the triple bark of the male shatter the silence of the night as they stay in contact and assess the locations both of members of their group and any neighbouring animals.

Pregnancy lasts 53 days, during which time the vixen will select and clean a number of den sites or 'earths' in which to raise her cubs. The chosen fox earth may be either self-excavated or an enlarged and disused rabbit warren or badger sett in the countryside, with favoured locations in the suburbs commonly being under garden sheds. Four to five cubs are usually born blind and deaf in mid- to late March. For the first two weeks, they will be constantly supplied with milk and attended to by the vixen; she, in turn, will be kept fed by regular provisions brought to the earth by the dog fox. When the cubs' eyes and ears finally open

they begin to stray much more until, after four weeks, they will eventually emerge blinking into the daylight as dark-chocolate-brown coloured fur balls.

Undoubtedly the best time to see the foxes' social and playful side is the period between their emergence and the time when the cubs have to stand on their own four feet in the autumn. Initially they will then remain close to their earth, playing and engaging in mock fights. While they look like they don't have a care in the world, these tussles are used to develop a social hierarchy and hone their hunting techniques, skills that could make the difference between life and death. As the cubs mature, they begin to spend their entire time above ground; they moult into their orangey-red fur, and their ears and snout elongate to produce the characteristic foxy appearance. The cubs are fed by their parents or other group adults at rendezvous points close to the den sites right up to July, by which time they will have started to hunt themselves.

The adults give the cubs very little training, so, initially, they are dependent mostly on easily caught food such as earthworms, beetles and small fledgling birds; if July is wet, more cubs will survive through to autumn as the worms will be easily accessible. As cubs begin to forage further away from the earth, their inexperience makes them vulnerable to predators such as other foxes, badgers, dogs and, of course, cars, so, where possible, they will try and use the centre of their parents' territory where they feel most secure.

During each breeding season around 425,000 cubs are born, and, as the fox population remains fairly constant, this means that as few as four in ten cubs make it through to the following breeding season to replace the older animals. This could mean the average life expectancy of a British fox may be no more than a paltry 18 months. After being maligned in the countryside, where it does not get credit for keeping rabbit numbers in check, and undeservedly blamed for taking pets in the towns, is it now time to cut the fox some slack? Its resilience, adaptability and endurance in the face of an ever-changing Britain shows that, as a species, it has more in common with us humans than we dare to think.

Fox cubs playing. This is integral to honing their hunting techniques and sorting out a pecking order.

36 Trapping moths

While butterflies are popular, iconic and colourful daytime insects familiar to everyone, moths have something of a PR problem. There seems a widespread misconception that moths are dull, boring and brown, and put on this planet to do little else than to chew our clothes and carpets. While a tiny minority do unfortunately have this tendency, and many can be brown, they are certainly not dull and boring as any moth trap will illustrate.

The differences between moths and butterflies are numerous, complex, indistinct, and include frequent exceptions. The most obvious difference is that butterflies fly during the day, while the vast majority of moths are either crepuscular – flying at twilight – or nocturnal by nature. A close look at the antennae of the two groups will also reveal that most butterflies have slender antennae with a club on the end; moths either have feathery antennae or a pair of simple filamentous strands without clubs. In resting state, all of the butterflies, with the exception of the skippers, close their wings over their backs; moths, with the exception of the thorns, lay their wings alongside their bodies. Most moths also possess a frenulum, which is a small hook on their hindwings that attaches to barbs on the forewing, whereas the four wings of butterflies all operate independently. Finally, most moths tend to have hairy or furry-looking bodies, with larger scales on the wings to enable them to conserve heat at night, while the sun-loving butterflies do not need this extra insulation so have more slender thoraxes and abdomens.

The angle shades looking less like a pair of sunglasses and more like a crumpled leaf.

Moths are also sub-divided into two groups called 'macro-' and 'micro-' moths based on their anatomical structure, but, as a few micro-moths are larger than macro-moths, this division can be complicated to the untrained eye. Virtually all macro-moths can be distinguished by the patterning on, and the shape of, the wing, which makes identification easier; the micros, on the other hand, invariably and unfortunately need their pressed genitalia to be examined down a microscope. Macro-moths are also much more numerous than our 59 resident butterflies, with over 800 species recorded in Britain. During a good night's 'moth-ing' in the summer, the 'moth-er' can be rewarded with at least 100 species and possibly as many as 1,000 individuals from a specially designed moth trap.

That moths are attracted to artificial lights has been known ever since man made fire, hence the phrase 'like a moth to the flame', but the reasons why this happens are complex and still not fully understood. When watching how moths become attracted to a light or moth trap, it is very noticeable that many of the individuals appear to fly around the light in ever-decreasing circles, and the most common theory to explain this behaviour is that moths use a technique

Moths

WHEN
Between May and July for most resident species of hawkmoth

WHERE
The moth trap can be placed anywhere from the garden to a local nature reserve

of celestial navigation called 'transverse orientation'. By maintaining a constant angular trajectory to a bright celestial light such as the moon, the moths can fly in a straight line. As the moon is so far away the change in angle between the moth and the moon's rays will be negligible, the moon will always be in the upper part of their visual field and no lower than the horizon.

Evolutionarily speaking, human light sources have been around for such a short space of time that the moths have not yet evolved the ability to ignore the light pollution we create. As a moth-trap light may be so much stronger than that of the moon when the insects are close by, they become confused and instead use the artificial light for navigation. But, as this light is below the horizon and the angle of the moth to the light changes markedly after only a short distance, the moths will instinctively attempt to correct this by constantly turning towards the light, causing them to spiral down to the light until they either hit it or drop into the moth trap.

A quick glance at any guide to British moths will soon make the reader aware of the infinitely varied colours of moths, with a number of species being easily as coloured and possibly even more intricately patterned than the butterflies. Many moths have been given wonderful common names as well, such as mother shipton, peach blossom or brindled beauty, which certainly adds to the fun of moth identification. Of all the wonderfully different 17 families of macros in Britain, though, without question the most spectacular group regularly encountered in moth traps are the hawkmoths.

The hawkmoths comprise about 850 known species worldwide and are most heavily represented in the tropics. The family includes such well-known species as: the infamous death's-head hawkmoth that was immortalised in the film *Silence of the Lambs*; the day-flying hummingbird hawkmoth; as well as the long-tongued Darwin's hawkmoth, which was predicted to exist by the great man after Darwin found an orchid that could only be pollinated by a proboscis of over 20 centimetres in length! The hawkmoths are moderate to large in size, with a wingspan of between 30 and 125 millimetres, and are characterised by their rapid flight, long, narrow hawklike wings and a streamlined abdomen, all clearly adaptations for quick and sustained flying. In addition to some species being able to hover, the hawkmoths are thought to be some of the world's fastest flying insects, capable of travelling at 30 mph.

The larvae of hawkmoths are also larger than the vast majority of the other moth larvae, and can reach 80 to 120 millimetres long, with a surprisingly thick body and a horn clearly present on the eleventh segment, close to their rear end. The family's Latin name, *Sphingidae*, comes from the posture adopted by the caterpillars: when they are resting on a twig, they cling to the plant with their pro-legs and hind claspers, while the front halves of their body rear up with the head curved back towards the twig, resulting in the caterpillars' profile resembling that of an Egyptian sphinx.

The caterpillars have an incredible appetite and their weight before pupation may be ten thousand times that of when they initially hatched from their eggs; this means that, during their six-week larval stage, these 'eating

The elephant hawkmoth derives its name from the grey, trunklike caterpillar, not from the startling pink and olive livery of the adult moth.

If disturbed, the
eyed hawkmoth will
flash 'the eyes' on its
hind-wings to give
the impression that
it is much bigger
and scarier than it
really is!

machines' will have to moult at least three or four times. Although some hawkmoths are considerable pests to crops, such as the tobacco hornworm (hawkmoth) in the tobacco-growing areas of the USA, all of the British species confine their voracious appetites to a mostly abundant range of native plants.

Some of the caterpillars, such as the poplar and eyed hawkmoths, are well camouflaged against their respective food-plants, with disruptive patterns making them difficult to pick out. However, a number of other species that feed on low-growing plants, like the immigrant spurge and bedstraw hawkmoths, often have conspicuous colours, presumably to act as a warning to birds of their distasteful nature, while elephant hawkmoth larvae have eyespots to startle would-be predators. Those caterpillars that avoid being predated will then bury themselves underground to pupate in the soil over winter.

In Britain there are nine breeding species, with a further eight that have migrated from continental Europe occasionally recorded at moth traps. The elephant hawkmoth is certainly one of those species that puts paid to the theory that all moths are brown, with its brazen pink and olive colouration. This small species is one of the most common hawkmoths in Britain and is widespread in England, Wales and southern Scotland. It is particularly common in urban areas, as its main food-plant is rosebay willowherb – that colourful coloniser of car parks, railway sidings and roadsides. The caterpillar is green or brown, speckled with grey, and has a pair of colourful eyespots on the fourth and fifth segments either side of the body, which become dilated when it is disturbed. The caterpillar's body immediately behind the head is long and extendable and, when the caterpillar waves its head around while looking for the next meal, it has more than a passing resemblance to an elephant's trunk, hence the name!

The hawkmoth most likely to be found as far north as Scotland, and even reaching the Arctic circle, is the poplar hawkmoth. The favoured habitats for this moth are any woodland margins, parks and gardens where the caterpillar's food-plants of poplar, aspen or willow can be found. The adult has quite broad forewings that are coloured delicate shades of grey, have heavily scalloped trailing edges and a distinctive rusty-red patch at their base. When this quite large moth is at rest, it is very noticeable that the leading edges of both smaller hindwings poke out in front of the forewings – the frenulum is absent in this species – and the tip of the abdomen also commonly curls upwards.

Of course the holy grail of hawkmoth finds must be that of the death's-head hawkmoth. The name of this most spectacular moth arises from the markings on its thorax, which bear a striking resemblance to that of a human skull and are complemented by the lateral stripes on the abdomen, which add a set of ribs to the image. This moth is the largest species regularly encountered in Britain and has been regarded in many countries as an omen of disaster, a myth possibly perpetuated by being the only known moth species to make an audible squeak when touched! The death's-head is a native of Africa and migrates northwards into Europe each year, often reaching mostly southern Britain in small numbers by early autumn. The preferred food-plant of the caterpillar is the potato, but they will also help themselves to woody nightshade and jasmine. This is most certainly one species that any moth trapper would be delighted to hear going 'bump in the night'!

35 Grey seals pupping

The autumnal months of September to November are not usually considered the traditional or best time to spend a day down at the beach or on an island-hopping sojourn. But this is exactly the time to visit one of a few special coastal locations if you want to catch up with grey seals as they pup – a tremendous wildlife spectacle that is one of our best-kept secrets and of which Britain as an internationally important player should be justifiably proud.

The grey seal is only one of two species of seal that breed along the British coastline, the other being the common seal, which, ironically, is the rarer of the two in Britain's waters. The wildlife novice can encounter difficulty in identifying these two superficially similar species, but it helps if you know that the grey seal's wonderful Latin name *Halichoerus grypus* translates as 'Hook-nosed sea-pig'! While hardly flattering, this translation goes some way to describing the looks of the greys: their elongated muzzle contrasts with the 'snub-nosed' appearance of the more diminutive common seals. Another vital distinguishing feature is the nostrils: in the grey seals the closed nostrils appear like parallel slits; in the common they are splayed to form a V shape.

The size and shape of the grey seal differs between the sexes too. The adult males or bulls are much larger and heavier than the females, with a massive pair of shoulders where the skin over this region and the chest consists of heavily scarred folds and wrinkles. Bulls have a distinctive and convex snout, giving them a 'Roman nose' profile, while the females, or cows, look more slender and seal-like by comparison. Although 'grey' is a reasonably accurate colour rendition of most of the seals, there is much variation, with some bulls being virtually black, while some cows are creamy-white with a few dark blotches.

It is thought that the grey seal's historic distribution was in a broad swathe across the entire northern Atlantic, but the advance of the ice during the last Ice Age 20,000 years ago, split the seals into a western and eastern stock. The western Atlantic population can be found along the coast and islands of Canada's eastern maritime provinces, and, apart from a small relict population in the Baltic Sea, the convoluted British coastline is thought to hold the vast majority of the eastern stock, with current numbers estimated at around 124,000 seals or 40 per cent of the world's population.

The breeding and mating of grey seals is very much an autumnal phenomenon, and the species is unique in being the only seal that lives in a seasonal environment yet produces its young at a time when the newly independent pups will have to contend with winter storms. It is possible that, being a species that is very vulnerable on the breeding ground, the grey seals have changed their breeding season comparatively recently as a response to

prehistoric predation. The peak of breeding also occurs in different months
in different parts of the range: pups are born in early September in southwest
England and south Wales; from late September to early October in western
Scotland; and November at sites on the east coast.

The breeding season begins with the arrival of the sleek, fat females that
have spent the summer fattening up on a large variety of fish, of which sand eel,
cod and Dover sole are thought to feature heavily. Favoured breeding locations
tend to be isolated and uninhabited islands with smooth, sandy beaches; a few
choice mainland beaches are also regularly used. In areas with a lower breeding
density of grey seals, such as in the southwest of England, caves are often used
for giving birth and mating. Close to the time of giving birth, the cows will haul
themselves to favoured pupping spots usually about a day before giving birth.
These spots vary from the beach just above the high-tide line to a location
several hundred metres from the water, such as grassy dune slacks.

Grey seal labour is not very obvious; the birth of the 14-kilogram pup can
often occur rapidly, with the first sign being the sight and sound of a host of
wheeling gulls as they squabble among themselves for the membranes that
enclosed the pup and the afterbirth. Immediately after the birth the cow will
sniff and touch the pup a number of times to learn its smell.

Seal pups

WHEN
From early September to
November depending on the
location, with southern and
western sites pupping first

WHERE
Monach Isles, Outer Hebrides;
Orkney Islands; Donna Nook,
Lincolnshire

A pup starts to moult into its adult coat in preparation for a chilly life at sea.

The most startling thing about the newborn pup is its creamy-white fur coat, which is thought to reveal the British grey seal's ice-breeding ancestry. Further north in the greys' range, many of the cows still give birth on the ice, meaning the pups are perfectly camouflaged, but, since the retreat of the ice from the British Isles after the last Ice Age, the pups have not been pressurised by predators to change colour and so stand out like a sore thumbs. In a technique still designed to offset predation by swamping the predators, the vast majority of females give birth with a remarkable synchrony, with the result that, on a dense grey seal pupping beach, the white pups can be seen regularly studded along the entire length of the beach with their mothers in attendance.

Initially upon birth, the pups are very poorly coordinated and would suffer quickly from the cold if they were to enter the water inadvertently. The cows stay close to the pups and react aggressively towards any other seals or gulls that come too close, and within a few minutes the pup will begin trying to suckle. Sometimes it takes a while for the newborn pup to locate its mother's nipples, as they are set towards the tail-end of the body and are usually inverted to aid the seal's streamlining in the water. However, some gentle prodding from the pup's muzzle causes the nipples to pop up and it then latches on with its specially indented tongue and begins to suck. Feeding bouts last no longer than about ten minutes on average, and the pup is fed at five- to six-hourly intervals on the incredibly rich milk, which is 60 per cent fat and resembles mayonnaise in consistency. The growth of the pup is slow for the first day but then increases at a phenomenal rate so that the pup will have doubled its weight in the first week. By the end of the lactation period, which lasts between 16 and 21 days, the pup will have increased by an incredible 3 times its birth weight. As the cows feed their voracious pups, they will begin to lose weight in a manner that would have many human mothers green with envy; they lose their rotund shape by shedding four kilograms a day during this intense period.

The bulls generally come ashore when the first pups are born and spend the lactating period competing with the other males for sole access in among groups of breeding females. Like the females, the males do not feed during the breeding season and live off the stores of fat that have been laid down at sea over the previous ten months, meaning that they can devote their entire attention to garnering as many females as possible. For the successful grey seal bulls, size matters, with the largest males, or 'beach-masters', often retaining as many as ten cows in their harems. Relations between neighbouring males are often amicable, but they can also result in brutal fights when two evenly matched individuals clash over ownership of the females.

Towards the end of the lactation period, the cows come into season and are then mated with their respective bull a number of times prior to the females leaving the beach and abandoning the pups to their fate. The gestation period of grey seals is only seven months, yet the cows will not be giving birth until they come ashore the following breeding season a full ten months later. They use a combination of suspended development of the fertilised egg and delayed implantation to ensure that the birth will be synchronised to take place at the same pupping beach the following year. Very little is known about where the seals spend their time once away from the breeding ground and they may roam far and wide in order to find the five kilograms of food a day they need to get through the winter.

At around three weeks and just as their mothers begin to leave them, the pups will shed their white fur to reveal the sleek dark coat needed to survive in the sea. It is no surprise that the first few months of life are the most dangerous for grey seal pups, with mortality reaching as high as 40 per cent on some crowded beaches where they can easily starve or be squashed by a careless bull. The process of learning to feed out at sea without any parental help inflicts further casualties, meaning that two-thirds of grey seal pups never reach their first birthday. After this first year, however, their chances of survival become much better, with the cows reaching sexual maturity at around four and living anywhere up to the grand old age of 25 years old. Male seals often do not reach sexual maturity until six, but it is thought that they are not able to hold a position in a breeding group until at least ten, and, even then, will only have a few years at the top before their physical decline leads to them becoming marginalised by the new generation of younger and stronger bulls.

Prior to the Grey Seal Act of 1914, seals were so heavily hunted that numbers in the British Isles were thought to be as low as 500 individuals, but thanks to this and subsequent laws, the population has boomed to a current all-time high of around 124,000 grey seals. While an unmitigated success story for conservationists, grey seals are a contentious issue for fisherman, who see them as competition for fish stocks.

High-tide roost

WHEN
August to November,
although waders are present
all winter

WHERE
The Wash; Dee Estuary;
Morecambe Bay; Thames
Estuary; Exe Estuary

34 Winter high-tide roost

The majority of Britain's highly convoluted coastline that is not made up of either rock or shingle consists of mud, glorious mud. Everyone loves a sandy beach and mudflats are often regarded both with distaste and as a waste of valuable space, but these huge, windswept and seemingly barren areas possess some of the most biologically fertile land we possess. They also play host during the autumn and winter to a huge influx of waders arriving from all points on the globe that take advantage of the muddy microscopic plant and animal life on offer.

It has been estimated that the 155 estuaries around the British coast may well comprise close to a third of the intertidal mud and salt marsh represented in the whole of Europe. These intertidal estuaries are so vital to overwintering waders that they are thought to hold a staggering 1.7 million waders during the winter months, with many more using Britain's muddy coasts as a refuelling pit stop before travelling on to overwinter elsewhere. This huge figure includes internationally important numbers of knot, dunlin, oystercatcher, redshank, curlew, bar- and black-tailed godwits and grey plover. Many of these species have originated from breeding areas as far away as northern Europe, Siberia, Iceland, Greenland and northeast Canada.

It seems that Britain has become such a 'winter wader wonderland' because of our combination of relatively mild winters and large tidal ranges, which ensure that extensive areas of intertidal mudflats become exposed for feeding on a daily basis. Mudflats become formed in estuaries when the discharging rivers slow to such a pace that they lack the energy to transport their cargo of silt any further, causing it to be dumped. In certain estuaries around the British coast, mud has been continually washed down and deposited since the end of the last glaciation, leading it to reach depths of as much as 30 metres. But, irrespective of the depth of the sediment, it is only the surface layers, which are in contact with light, water and air, that support the chain of life from bacteria to algae, and from worms to shellfish, that are ultimately exploited by the birds.

All mudflats undergo the experience of tidal inundation twice every 24 hours as the tide rises and falls, and it is this tidal rhythm that governs the behaviour of all the creatures that live either in or on the mud. The intertidal zone is a harsh place in which to live, as the coming and going of the seawater creates an environment of constant change, resulting in hugely varying temperatures and salinities. Many plants also find it difficult to establish a foothold on the fine shifting sediments; this results in much of the plant material that occurs on the intertidal mud, such as diatoms and algae, being of a microscopic size. These factors also mean that relatively few invertebrates have been able to make a permanent home in the mud compared to those in more stable habitats. But

The retreating tide reveals a smorgasbord of food for the hungry knot.

those that do manage to become specialised enough to cope can reach vast numbers because of a lack of competition for the organic material available and the huge amount of substrate (mud) available in which to live and from which to feed.

In addition to mud being easy to burrow into, its structure of fine particles enables the construction of temporary or permanent burrows for lugworms and ragworms, and a variety of shells, such as Baltic tellins and common cockles, in which they can conceal themselves while probing above the surface into the tidal water to feed with specialised siphons and filters. Other creatures, such as *Hydrobia* snails and *Corophium* sandhoppers, also emerge from the mud at certain times to graze on the organic debris at the surface; they do so in such profusion that thousands may be present in each square metre of mud.

The best way to appreciate the vast numbers of living creatures hidden in the mud is to watch the multitude of waders that descend to feed on them. These wader flocks will often contain a whole range of species feeding side by side with little or no aggression, as their techniques for foraging are so variable that they will rarely come into competition with each other. As the wader species have no objection to flocking together while feeding, they are able to gain all the benefits of being in a crowd, such as greater security from predation and the ability to exploit patchily distributed food with greater efficiency by noting 'feeding hotspots' where other waders are feeding successfully.

Waders are thought to find food either by touch, sight, or, in some species, by a combination of the two senses. A number of species take advantage of the fact that invertebrates become more active as they are covered by the advancing tide, so there is often a concentration of birds along the tide edge. Knot feeding here are believed to forage primarily by touch as their bills seek out the

With no mud exposed at high tide, the knot and oystercatchers avail themselves of the chance to catch up on forty winks.

microscopic snails and bivalves just below the surface, while godwits visually search either below water or at the tidal edge for the recent casts of lugworms before using their long sensitive bill both to feel for, and extract, the worm. In addition to the birds working the tideline, a number of waders have become specialised at feeding in shallow standing water; these include avocets, which swish their scything bills underwater, and spotted redshank, which upend themselves in a manner more reminiscent of dabbling ducks than waders.

Ahead of the advancing tide, or behind one that is receding, many birds prefer to feed in damp mud because it is easier to probe. The majority of sandpipers, such as redshank or dunlin, feed by scanning the mud to either side of them in walking transects, and, when potential prey is spotted, capturing it either by a peck, if it is a small *Corophium* snail, for example, or by probing more deeply if it is a buried clamworm such as *Nereis*. Short-billed birds like grey plovers and lapwings do not tend to use the 'touchy feely' techniques of the sandpipers, and rely more on their excellent sight to scan around them for signs of activity while they remain rooted to the spot. When they catch sight of a prey item worthy of capture, they quickly sprint to the spot before it burrows out of their depth. This technique has the advantage that the mud is less disturbed and so does not form a localised 'prey depression', which will often happen when the sandpipers are feeding like a miniature herd of grazing wildebeest. When prey activity is low, however, the plovers are not able to begin feeding by touch like the sandpipers, in which case they can resort to a different foraging tactic known as 'foot trembling'. This technique involves placing one foot on the sediment slightly ahead of the other and rapidly vibrating it up and down, which, it is thought, is meant to simulate the approaching tide, causing their prey to come prematurely to the surface to feed, thereby fatally revealing their location.

Bill length is, of course, very important for determining which birds are able to reach various prey items at different depths, with the curlews' long sensitive

LEFT: The diminutive dunlin contemplating where to probe its bill next.

RIGHT: A feeding bar-tailed godwit – with worm.

'Birds of a feather
flock together',
particularly if you are
a knot at high tide.

bill being the obvious tool that equips them well to winkle out even the most
deeply buried lugworms and ragworms. While birds like dunlin and sanderling
are not able to penetrate the mud to the depths of which the curlew is capable,
it is thought that they may well possess the ability to use sensors in their bills
either to smell or taste the location of prey hotspots.

For most species of intertidal foragers, the usual peaks of feeding activity
are on the ebbing and flowing tides, with something of a lull around low tide

when many of the birds will roost for a while close to their feeding grounds. Feeding time for these birds is governed strictly by the availability of mud so, irrespective of whether it is day or night, they must feed when the tide is out. While feeding at night means there is less light in which the waders are able to spot their prey, this is compensated for by the fact that the invertebrates tend to be more active meaning that moonlit nights can often represent very rich pickings even for the plovers that hunt mostly by sight.

As the turning tide begins to flood in and cover the mud, the different species will be forced to crowd together as the mudflat steadily diminishes in size. The birds usually seem reluctant to leave these feeding areas for the high-tide roost, resulting in them often overlapping one another in a rippling Mexican wave effect up the estuary as progressively more mud becomes covered by the advancing tide. It is only when the last of the mud finally becomes covered that they are grudgingly forced to take to the air in a mass of whirring wings and amid much noise. When the birds rise as one it can be a spectacular sight, particularly if there are large numbers of knot present, like at Snettisham in Norfolk, as they pack very tightly together to form a huge, swirling smoke cloud for the, often short, hop across to their high-tide roost on the nearby gravel pits.

Waders choose their high-tide roost sites very carefully and tend to prefer sites that have a good all-round visibility, freedom from disturbance and shelter from the wind. The choice of roost usually ends up being a compromise, of course, as the most sheltered sites will invariably have poor visibility, but the favoured locations are usually tried and tested spots such as nearby fields, salt marsh or a remote section of beach above the tideline. Once the roosting birds finally settle, they adopt a posture to minimise heat loss, such as facing into the wind in order to avoid their feathers being ruffled, shortening their neck, tucking their bill under the feathers and standing on one leg. While the birds will undoubtedly sleep for short periods in the roost, they will often keep one eye open and alert for any potential danger. Should a peregrine approach the high-tide roost, for example, the waders will immediately assume an alert posture, with their necks extended and their wings held slightly away from their bodies so they can be ready to take to the air in evasive action. When the waders are forced to roost at night this can present a different problem, as they suddenly become more vulnerable to stealthy ground attacks; in this scenario they will often prefer to roost in shallow water to deter any terrestrial predators, such as foxes, allowing them see the sun rise over the mudflats another day.

Machair in summer

The word 'machair' comes from Gaelic and means an extensive low-lying and fertile plain. The term for this habitat encompasses everything from their white sandy beaches, to the calcium-rich dune pasture, to where the sand encroaches on to the peatlands further inland. Taking a walk along the machair in spring when the birds are calling, or in summer when there is the most incredible blaze of wild flowers is like taking a step back to the halcyon pre-industrialised days of farming on mainland Britain.

With its precise requirements for formation and its localised nature, machair is one of the rarest habitats in Europe. In the British Isles, it is found only in the north and west of Scotland and western Ireland, of which almost half occurs along virtually the entire western fringe of the Outer Hebrides. The machair habitat as we know it today was formed at the end of the last Ice age after the melt water from the glaciers deposited enormous quantities of sand and gravel into the sea over what is now the Continental shelf. As the sea level rose, this glacial sediment, which became mixed with the crushed shells from marine molluscs, was then driven ashore by waves from the strong prevailing south-westerly winds to form the characteristic white beaches. Over time a constant supply of this sand caused some to be blown above the high-tide mark and began the formation of dunes. Centuries of constant wind has occasionally broken down these dunes, depositing this fine white sand on to the fields and pastures beyond, even blowing far enough inland to coat some of the peat bogs.

As machair sand is composed of 80 to 90 per cent crushed shells, these western beaches are white in colour, as opposed to the more typically yellow-coloured beaches on the eastern coasts of outer Hebridean islands, such as North Uist, where the sand is mostly from mineral-based material. Down below the sea line on beaches with a westerly aspect, the sand is repeatedly exposed to the action of the waves and wind meaning it is a highly mobile environment and devoid of any colonising plants, but just above the high-tide line the first few hardy, pioneering plants like sea rocket and sea sandwort begin to take hold. The very presence of these plants initiates the eventual formation of dunes: they provide a barrier to sand particles which then become deposited, and in turn create a bigger obstacle as more sand becomes accumulated by the plants as it is blown up the beach. The number one dune builder has to be marram grass, as its spiky inwardly rolled leaves, rapid growth rate, tussocky nature and deep root system mean that it can thrive in this harsh, sandblasted environment.

Behind these dunes – which can reach up to ten metres in height – the impact of the wind and salt spray are much reduced, meaning that a larger variety of plants are able to grow in the bare sand. As these plants decay over the seasons, the embryonic soil holds moisture a little more easily and the

The floral extravaganza that is the machair.

Machair

WHEN
Late May to August

WHERE
West coast of the Outer Hebrides, Tiree, Coll and other small Scottish beaches with a westerly aspect

alkalinity is slightly lowered, meaning that plants such as butter-cups and lady's bedstraw may be able to take hold, eventually forming meadows as the sand becomes habitable.

From the end of autumn until at least the middle of May, the machair has been described as a 'desolate waste of sand', but equally large areas of the machair can be flooded. This serves to protect these vulnerable grasslands from wind erosion and to provide rich feeding grounds for wintering wildfowl such as barnacle and Greenland white-fronted geese. Atlantic winter storms wash up a huge amount of kelp from just offshore and this forms an additional sea wall along the dune edge, which helps to protect the machair from being inundated by sea water. This kelp, or 'tangle', has been collected by the local population ever since Neolithic times as a natural fertiliser and 40 tonnes per hectare is still placed over the machair once it has drained each spring to improve the organic content. The fields are then ploughed to help bind the soil together and to improve its moisture-holding capacity, which in turn makes the grassland more resistant to wind erosion and ready for planting crops.

The corncrake, a.k.a. *Crex crex*. So repetitive they named it twice!

Spring generally arrives late on the machair because of cold easterly winds, but the application of the tangle, the ploughing, and careful grazing before the main growing season help create the perfect conditions for a vast array of wild flowers to grow alongside the planted oats, rye or barley crops. Strict field rotation is practised and, in some of the fallow areas, the floral diversity is so rich that it can reach an astonishing 45 species per square metre.

Early in the season the white confetti of daisies cover huge areas of machair; come June, the machair turns yellow as buttercups and bird's-foot trefoil dominate in the drier areas, and silverweed, yellow rattle and marsh marigold thrive on the slightly wetter ground. Later in the summer the predominant colour tends to move to the red and purple end of the spectrum with red clover, ragged robin, self-heal and field scabious taking centre stage. The machair is also famous for its orchids, with pyramidal and fragrant orchids occurring in profusion alongside the unique Scottish marsh orchid at a number of sites on North Uist. The application of very low concentrations of herbicides means that, growing among the crops to be harvested, agricultural weeds, such as corn marigold and charlock that are all but extinct on the mainland, can be found.

This phenomenal floral diversity provides food and accommodation for a wide variety of invertebrates such as snails, grasshoppers, spiders, harvestmen and rare bumblebees, which, in turn, provide food for a range of agricultural birds that have declined massively on mainland Britain but of which the Outer Hebrides still have healthy populations. The machair and adjacent crofting lands on the Inner and Outer Hebrides have now become the last remaining strongholds of the corncrake, with recent surveys indicating that the islands may well hold at least 90 per cent of the British population of just under 600 calling males. This notorious skulker was difficult to see even in its heyday, when it could be heard repeating its strange rasping call up to 20,000 times a night all over rural Britain. Looking rather like a cross between a grey and rusty-coloured chicken and a moorhen, the corncrake began to decline on mainland Britain at the turn of the 20th century because of agricultural intensification. As hay fields were cut mechanically, the grass was cut earlier and removed for silage in early summer, the nests, young and even the adults disappeared rapidly. However, late cutting and the low-level of mechanisation is still commonplace in the Outer Hebrides today, which allows the corncrakes to hide in plentiful cover at harvest time, giving them a fighting chance. Grants and subsidies in return for good farm practice are also now available, and corncrake numbers seem to have started very slowly improving on the islands.

Other birds present in healthy numbers on the machair, while becoming rarer on the mainland, include corn bunting and twite. The lack of pesticides used on the crops means that there are plenty of beetles and caterpillars that birds can catch to feed their young. Also, after the harvest, the abundant fields of stubble and seeds from wild flowers ensure there is enough food for the birds to survive the winter as they rove the machair in their large flocks.

Of all the birds on the machair, the habitat is most famous for its breeding waders, with an estimated 17,000 pairs on the western fringes of the Uists and Barra alone. The most numerous breeding wader is the lapwing, but there are also large numbers of dunlin, ringed plover, oystercatcher, redshank and snipe. Because of the rich mosaic of habitats, the rotational form of agriculture and the lack of ground-based predators on the islands, these machair locations may well hold close to 40 per cent of the entire British breeding population of dunlin and close to a third of lapwing and ringed plover.

Curiously, the only recent threat to the waders has been the introduction of the hedgehog, as Miss Tiggywinkle has an unfortunate penchant for wader eggs. With efforts being made to tackle this problem, however, there is hope that the display calls of the lapwings, snipe and corncrake, in addition to the vast wild flower blooms, will be spectacles for many years to come in the land of the machair.

32 Wild goats rutting

While watching wild goats negotiate the most terrifying vertical cliffs and sheer precipices with such adroitness in places like Snowdonia, it is hard to believe that these hardy beasts only exist in some of our wildest places thanks to a helping hand from prehistoric man. These descendants of the original domesticated goat stock look as much part of the scenery as the drystone walls that weave up and down the mountainsides.

Feral goats in Britain are confined to mountainous districts, cliff tops and islands; they are widespread in Scotland, and occur locally in a few remote locations across northern England and Wales. Perhaps the best-known and most easily encountered population is the famous feral goats of Snowdonia. With the red deer long since exterminated from Wales's most famous national park, it is curious that the only remaining large herbivore that is both tough and canny enough to survive in this unforgiving landscape, is an introduced goat.

It is thought that goats were originally introduced to North Wales when Neolithic man first crossed the Channel to colonise Britain from mainland Europe over 5,000 years ago, bringing his domesticated livestock, such as goats from the Middle East, with him. Then, around 1,500 years ago, during the Iron Age, the farmers are believed to have taken their goats into the mountains to feed and the animals' descendants are thought to have stayed ever since.

For many centuries the goats held sway as they were used for their hair, hide, milk and meat; even up to the Middle Ages, goats were believed to have been more abundant than sheep, as they were able to graze the precipitous crags to the exclusion of less sure-footed and often more valuable cattle. It was not really until the 19th century that goat numbers began to decline as sheep numbers began to rise, due in part to the high demand for wool. By this time, however, many of the goats had become feral and the wild population was bolstered by escaped goats or those that were let loose once no longer needed.

The wild Snowdonia goats of today most closely resemble breeds that have not been seen in domestic herds for over 100 years, meaning they have considerable historical and cultural value, in addition to being an integral part of the local wildlife. The goats are in small clusters around the park, with well-known herds existing on the Glyders, around Beddgelert, on the Moelwyns and the Rhinogs; since these groups are largely isolated, this has resulted in subtle differences in appearance between the herds.

The feral male goat or billy is an imposing sight, reaching 90 centimetres at the shoulder and weighing between 45 and 55 kilograms, with the main difference between the wild version and the domesticated goat being that the wild goat has much longer hair, enabling it to tough out the worst possible Welsh weather. The females, or nannies, are much smaller, attaining a height

Since having been introduced, goats now seem part of the furniture in some of our wildest places.

Wild goats

WHEN
October and November for
the rut, although the goats
will be resident all year round

WHERE
Snowdonia, North Wales;
Lundy Island; Valley of the
Rocks, Lynton, Devon

of no more than 70 centimetres at the shoulder, with shorter hair and weighing approximately half the weight of their male counterparts. The coat colour of the males is generally piebald with black or brown blotches, but some can be entirely grey or black all over. The females, however, are usually much whiter and can be picked out among the rock scree at distance and can also be distinguished from the sheep as they only appear as pale as the goats when they are freshly shorn.

Apart from the obvious size difference, the sexes can be easily distinguished by their respective sets of horns: mature males have a large set of curled horns which can curve back towards the centre; the females' are thinner, straighter and more pointed. Both sexes grow rings around their horns (although they are more prominent on the male) with each ring representing a year's growth and the distance between the rings decreasing in the oldest goats. Unlike the antlers of red deer, which are shed after the rut and then regrown every summer, if the goat breaks its horns they will not grow back. Males that have lost their horns generally compensate of being heavier than the horned billies of an equivalent age. Both sexes also have the celebrated 'goatee' beards, although the tuft on the males is usually not to be as visible within their long coat.

The goats in Snowdonia tend to stay on the high ground most of the year, only descending during really harsh weather. The females spend most of their time in small groups of three to six within the boundaries of their home range, while the males will wander larger distances so they can visit and monitor several female groups. However, during the breeding season in September and October, much larger aggregations will form, making this the best time to look for these naturally wary animals as they drop their guard slightly while mating.

Brief fights are common between the males as they jostle for the right to mate with the nannies, but the most prolonged scraps occur when two males of a similar age and horn size meet hoof to hoof, with the reward of a group of receptive nannies going to the winner. In a scene more reminiscent of primeval fighting ibex in the Alps, the two evenly matched billies will then rise up on their rear legs before clashing their horns together with full force a number of times. In addition to the head-butting, they will also engage in wrestling and pushing matches more commonly seen among red stags, as each billy tries to assert control and gain the all-important mating rights.

After mating, the goats will wander much less during winter as they devote their attentions to searching out food, with their multi-purpose horns sometimes being used to scrape away snow to access the vegetation below. The goats will of course know their home ranges well and, when the weather is particularly inclement, they may well rest in favourite retreats in among the crags or scree slopes until conditions improve. The female groups stay together until the end of February, when the mothers will peel away to give birth to a single kid of two to three kilograms in among the rough scree or rocky terrain.

The kid is left hidden for the first couple of weeks, with the mother coming back several times during the day and night to suckle it, after which the kid joins the mother in the all-female group, where the two will keep in close contact by

For good or bad, these hugely adaptable and resilient animals seem here to stay.

call until the youngster becomes weaned. It seems that climbing in the most precipitous areas is an inherent trait for goats and, in no time, the kid will have taken to the cliffs and the scree slopes like a duck to water. Very occasionally twins are born, but, in such a demanding environment and with the mother only able to produce a certain amount of milk, it is very rare for both to survive.

The number of goats within the Snowdonia National Park has increased over the last 30 years from a relatively stable number of between 250 and 320 to around 500 goats, making them easier to track down now than for a generation. It is thought that this increase may well be due to the recent run of mild winters, and, with climate change affecting mountainous environments perhaps more strongly than elsewhere, their numbers are predicted to increase, causing a knock-on effect of too much grazing unless the population is kept in check.

Contrary to popular opinion, there seems little competition between goats and sheep for food, as the goats predominantly graze on woody, shrubby and coarse-leaved plants such as the bark from trees and the leaves of heather and bilberry in areas that the sheep may not be able to reach. The goats, however, are capable of causing considerable damage to trees by bark stripping and browsing, and, where this has happened in some of the important woodland sites on the lower slopes, they have had to be immediately removed. Undoubtedly, the goats have also caused resentment among local farmers as they steal the mixed-feed left out for the sheep and damage drystone walls. All interested parties, therefore, agree that the numbers of goats will have to be carefully controlled in order to achieve a healthy and balanced population that will harmoniously coexist with, rather than overgraze, the environment.

While unquestionably non-native, the goats have a place in both the cultural heritage and ecology of some of our most treasured wild locations. So, if you enjoy both high-octane wildlife behaviour and being in the mountains, then the feral goats could run the Alpine ibex a very close second!

Swallows and martins

WHEN
Mid-April to late October

WHERE
Mostly open locations with safe nest sites, and preferably open water, anywhere in rural Britain

31 Swallows and martins feeding

The swallow is surely one of Britain's best-loved birds. After the depths of winter and the slow, cold wind-up to spring, the first swallow arriving back in Britain after its marathon journey from southern climes is a most welcome sight. The harbinger of summer was perhaps best summed up by the famous naturalist Gilbert White, who, in 1789, called swallows 'inoffensive, harmless, entertaining, social and useful'.

Well-fed swallows preparing for departure to sunnier southern climes for the winter.

The swallow is very much a bird of open habitat, such as farmland with pastures, or parkland with water close by, where they are able to hunt for insects; additionally, they like some form of building in which they are able to nest. The species has historically benefited enormously from the widespread clearance of forests by humans, and has long since abandoned its traditional cave or tree nest sites in favour of nesting alongside us. Having originally provided the perfect conditions for the swallow, our challenge is to keep their population healthy in the face of agricultural intensification and the general tidying up of the countryside.

The European or barn swallow is one of three species of hirundine – birds of the swallow family – that can be regularly encountered in a British summer, the other two being the house martin and sand martin. It is also the species with the largest distribution of any of the 75 hirundine species worldwide, with a total of six sub-species being found across five continents at some point during the year.

Despite the martins and the swift looking superficially similar, the swallow is immediately identifiable even to the novice birdwatcher who is able to find a few moments to watch and admire this handsome and aerobatic little bird. The swallow's uniform blue-black upper parts, which are burnished with a metallic sheen, contrast with the chestnut red of the bird's forehead, chin and throat, with a dark-blue pectoral band separating the darker face from the white plumage of its underparts. A characteristic feature of most hirundines is a forked tail and this is particularly the case with the male swallow, which has greatly elongated outer tail feathers that form very noticeable streamers while either in flight or perching. As befits an animal that needs excellent visual acuity to catch aerial insects, the swallow has its eyes set slightly more forward-pointing than most small birds. Its legs, however, are short, and, while well adapted for perching on telegraph wires or clinging to the nest, are not much use for walking.

Virtually the whole swallow and martin family is insectivorous, with their prey being taken in flight or occasionally from perches; the swallow, despite

House martins feeding in flight formation. With their white rumps showing, they bear more than a passing resemblance to mini killer whales!

having a small bill, has evolved a wide gape, which is further enhanced by specially modified feathers, called rictal bristles, around the beak that form the perfect catching apparatus. The swallow also has a large wing area in relation to its size making it extremely manoeuvrable in flight and able to twist and turn quickly to catch a particularly juicy bluebottle. It is also one of the smallest birds to master the glide, which it uses as an energy-saving device in between shallow wing beats. In contrast to the house martin and their distant relative, the swift, which are capable of feeding at height, the swallow invariably feeds at low level and skims the ground mostly below a couple of metres. When seen whizzing past, the swallow gives the impression of moving quite quickly, but in reality it is quite a slow flyer, using agility rather than speed to catch the aerial insects.

The swallow is a long-distance migrant over almost all of its range, with those individuals that breed furthest north generally travelling the furthest south for the non-breeding season. As British swallows over-winter in South Africa, the long return journey to breed is not without hazard, and there can be considerable fluctuations in the numbers of breeding birds arriving back to breed if unseasonal weather is encountered along the way. Unlike the sand martin, whose populations have seen large declines, the numbers of swallows is thought to have remained stable, with the most recent estimate being 726,000 pairs.

The swallows' annual cycle begins with their arrival in Britain. Turning up slightly later than the sand martin, which is usually the first migrant to hit the south coast in early March, the first swallows – usually the oldest and most experienced males – arrive a couple of weeks later. These birds do not waste any time; they head inland to their former breeding site, quite possibly to the very same nest they used the year before.

Young males that have not previously bred also return to their natal area as they attempt to find a nest site with which to attract a mate. Once the nest site is selected, the male will then advertise both his presence and his availability with display flights, while unleashing his trademark unmusical, twittering song above his chosen spot. When a female has been persuaded to check out the future residence he will court her with a special flight that involves flying around in slow circles around the nest site to seal the deal.

Unlike the sand martin, which nests colonially in sand or river banks, the swallow nests in a solitary fashion or in a loose-knit colony where suitable habitat and nesting opportunities are abundant, with outhouses or barns of particularly old-fashioned farms being considered the most desirable residences. Although swallows will nest in odd places such as the pockets of a farmer's overalls left hanging up, they will usually build enclosed mud nests on man-made structures such as beams, eaves or old shelves. The swallows generally prefer to nest in quite dark corners, where predation is much less likely; the only other essential prerequisite is that there must be continual access to the nesting site, with a small broken window pane or ill-fitting door proving no obstacle to these agile little birds.

The nest is constructed with mud from nearby puddles. This is carried back in the swallow's mouth before the little mud briquettes are deposited; the walls are reinforced by hay and hair to give added strength as the mud dries. It generally takes the pair between six and twelve days to build the nest from scratch, with mornings being the optimum time for building work and with plenty of rests in between. The female completes the final stages by lining the nest with feathers. Some nests are so well made that they will frequently last for decades. The same nest, however, is rarely used for second or third broods

With hungry mouths to feed, the adults need the long summer days to keep their brood satiated.

during the same year to avoid the build-up of blood-sucking parasites, which latch onto the chicks.

Egg laying commences soon after the nest has been finished and can be as early as late April in southern Britain or mid-June further north. The average first clutch is four to five eggs, with clutches further north tending to be larger, as swallows in the north will produce fewer broods, during their shorter summer. After around two weeks, when the chicks hatch, the rearing process seems very equitable between the sexes, with both parents sharing in the brooding of the young chicks and the task of feeding them. The flight feathers tend to begin appearing on the fifth day and the chicks usually fledge by the end of the third week, leaving the parents to breed again where possible.

Early in the season it is thought that 80 per cent of the swallow's food will consist of large flies, although they will often be forced to catch other insect groups during inclement weather. The composition of the insects during the course of the season, however, can change radically according to what is most abundant, and second broods are thought to be fed predominantly on greenfly and booklice, which are much smaller than the large flies, meaning the adults will have to work harder to get enough food. It has been estimated that a half-grown swallow brood may need around 6,000 insects a day, or 150,000 food items, to raise one nestling; an average brood will, therefore, demolish an astonishing half a million insects during a season. These insects are mostly caught within a few hundred metres of the nest site so older, traditional farms with insect-rich pastures often have better pickings than the insecticide-laden modern farms, where the swallows may be forced to hunt further afield.

The disappearance of swallows in the autumn and their miraculous reappearance the following spring led many scholars and naturalists from Pliny to Linnaeus to believe that swallows crept into the mud at the bottom of ponds. This falsehood was fuelled partly by the fact that large flocks of swallows temporarily use reed beds as a staging post before finally leaving for South Africa, and it was not until the end of 18th century that swallows were regularly seen out at sea and the basics of migration came to light. Unlike many birds, swallows delay their moult until they have made the journey to their winter quarters, with the majority of British swallows finally peeling away in late September and early October.

Sand martins are usually among our first migrants to return, with their primary job being to give their nest site a spring clean before breeding commences.

All the evidence seems to suggest that our swallows tend to return to the same areas, with the Cape Province at the southern tip of South Africa being the most common destination according to recoveries of ringed birds. It is thought their route takes in western France, across the Pyrenees, down the east coast of Spain and then across to Morocco. From there they travel southeasterly across Nigeria and Cameroon, through the Congo and Botswana to South Africa. The time taken for the journey down to South Africa can be around six weeks, but this can be spread over a couple of months as they move through southern Europe at a leisurely pace.

So, the swallow's grace, beauty and confiding nature, together with the fact that it eats large quantities of insects and lives alongside humans, mean that Gilbert White's assessment over 200 years ago was just about spot on.

Massed ranks of swallows on the move to South Africa.

Otters

WHEN
Any month, but more
commonly seen in spring and
summer

WHERE
Shapwick NNR (Natural
England), Somerset; Big
Waters Nature Reserve,
Northumberland (Newcastle
City Council)

30 Otters fishing

While an unhealthily large number of Britain's animals seem to have undergone a severe decline over the last 30 years, it is refreshing to find one member of our fauna that is staging a remarkable comeback. Although seeing a river otter may still be a pipe dream for many naturalists, all evidence suggests that we are more likely to come into contact with them now (or at least with evidence of their presence) than at any time since the 1950s, when they were still common enough to be considered fair game in an unfair match with otter hounds.

The coastal otters are more commonly seen during the day and, hence, are much easier to see than their rarer, shyer and largely nocturnal freshwater cousins.

The Eurasian otter is a very widespread species; it ranges from Britain in the west to Japan in the east and from North Africa and the forests of Indonesia in the south to northern Finland. However, a combination of factors mean it is now scarce or even absent from large parts of its original home range. The British population is one of the largest and most important populations in Europe. However, with the exception of parts of the west coast of Scotland and the northern islands of Orkney and Shetland, where the otters tend to be coastal and regularly seen during the day, it is still incredibly difficult to watch river otters in the wild because of their nocturnal and secretive nature.

The otter is one of Britain's largest carnivores and can be confused with little else. The male or dog otters can easily reach over a metre in length and eleven kilograms in weight; they are generally around thirty per cent bigger than the females. When seen, their flat head and broad muzzle, which is covered in long, stiff whiskers, can clearly be identified above the water, with the top of their back and occasionally their tail as they create a V-shaped wake in the water while doggy-paddling around, in between diving for food. Their long and thick tapered tail is one of the otter's most distinguishing features on land; combined with their short legs and long, sinuous body they have a silhouette that can only possibly be confused with the introduced and unwelcome American mink which has much darker fur, but is about half the size and a tenth of the weight.

The otter is perfectly adapted for an aquatic lifestyle, with its lithe body, webbed feet and muscular tail, which powers the otter while underwater. The otter is kept warm by a dense under layer of fine hair, which traps an insulating layer of air, and a thicker outer layer of much coarser guard hairs, which give the otter a very sleek and slick look when wet.

In a typical dive, otters will not spend any longer than 10 to 40 seconds below before resurfacing for a breath, and it is one of the great joys to follow an otter underwater by watching the trail of bubbles while it is completely absorbed in ducking and diving for its dinner. Since they are able to adjust the curvature of their eye lenses, otters are thought to be able to see as well underwater where there is sufficient light as above, but, for an otter feeding in

The V-shaped wake, the broad head and flat rudderlike tail ... congratulations, you are watching an otter!

murky water or in the dark, the sensitive bristles play a major part in tracking down their prey on the muddy bottom or along the banks.

As freshwater otter sightings are so rare, otter watching can be much more about looking for their signs than enjoying the otters themselves; they often leave behind visible clues as to their presence. Otter faeces, or 'spraints', are the most commonly encountered field signs and, when found, are definitive evidence that the stretch of coastline or river being surveyed is within an otter's home range. The spraint consists of all the undigested remains of their food, such as bones, fish scales, fur and feathers that have been bundled up and coated with mucus before being ejected. These deposits can vary from being a tarry-looking smear to a well-formed black pellet, but always have a pleasant sweet yet musky odour to them, which otter experts consider is vaguely reminiscent of jasmine. In addition to being just-voided bowel material, spraints also convey a whole range of information – such as the age, sex and breeding status of the owner – to other otters that may be passing through. These little messages are also placed in prominent positions such as on large rocks, fallen trunks, concrete ledges under bridges and otter entry and exit points along the stream or river. Higher concentrations of spraints also occur in rivers with a healthy population of otters.

The five-toed otter print is another very useful field character, particularly when the signs are encountered crossing soft mud, a sandbank or even snow. In addition to five toes appearing nearly in a straight line in front of the bean-shaped pad, claw marks and even the webbing can be distinguished between the toes on very clear prints. The width of an otter print varies between 5 and 7 centimetres, with any marks wider than 6.5 centimetres usually indicative of a dog otter. Care does need to be taken when counting the toes, however, to ensure that it is not the four-toed print of a passing dog or a fox.

Otters are capable of exploiting a wide range of aquatic environments from small ditches and streams, to large rivers, estuaries and the sea coasts, but for these habitats to hold otters they must provide the basic requirements of food and shelter. Locating otter holts, where a litter will be born and reared, and their resting places or couches can be very difficult. Couches are often situated above ground and may well be positioned under a pile of sticks, in the middle of an impenetrable bramble bush, among river debris or on scrub-covered little islands, while holts are most commonly located under the well-developed root systems of mature bank-side trees. The coastal otters in places like Shetland do not have the luxury of being able to find nooks and crannies under old trees so they dig their holt out of an area of peaty soil that has easy access to fresh water. This enables them to wash the marine salt – which would otherwise affect the insulating properties of their fur – out of their pelt regularly.

Both coastal and riverine otters are capable of travelling large distances in search of food, but most adults will stay within a well-defined area or home range that holds everything they are likely to need during the course of an otter's year. Unlike territories, which are often held exclusively by an animal or breeding pair, otter ranges tend to overlap and are linear features that consist of a number of miles of river bank or coastline. The length of these ranges varies enormously, with female coastal otters often joining forces to form groups, while the males possess much larger home ranges that may encompass several groups. In freshwater systems the otters tend to be much more solitary, with over 40-mile home ranges for the mobile males, through which they pass briefly only every couple of weeks, not being considered uncommon.

Coastal otters in places like Shetland tend to hunt during the day, with mornings and evenings being the best, as opposed to the riverine otters, which emerge around sunset to feed all night. When they are not inside their holt, the otters will spend at least half their time actively hunting, with the rest of their waking hours spent either grooming or resting. Fish are thought to make up 70 to 95 per cent of all otters' diets, but exactly what they eat is largely dictated by what is available; this can vary between location and throughout the year. Otters are primarily opportunistic: salmon, for example, will be caught at salmon rivers like the River Dart in Devon during their migration, while eels predominate on the Somerset Levels all year round, but with the number taken dropping during the summer as young birds represent an easy catch instead. At other sites, frogs and toads can be seasonally important and, in Mull, particularly young otters will take high numbers of crabs as they are easy to catch. The prize for innovation goes to the otters in Shetland, however, as they have even been seen catching rabbits in their burrows!

In England, it is thought that the river otters may well give birth at any time of the year, which differs from Shetland where most cubs are born during the summer months to capitalise on the much greater abundance of coastal fish at this time. The gestation period is only nine weeks and the young are born blind and helpless but fully furred, with two to three being the normal litter size in freshwater; this is very slightly lower for the coastal females. The development of the cubs is slow as they only open their eyes at four to five weeks. Even though they are weaned at three months, they are still very dependent on their mother to catch their food for them, only achieving independence at between seven and twelve months after their birth.

Otters are at the top of the food chain and, since the extinction of British wolves, they have nothing to fear apart from man and his dogs. Otter hunting dates back to King Henry II in 1175 and was particularly popular in the Elizabethan era, continuing right through to Victorian times and even into the

A sure-fire sign that an otter is in the vicinity – its very pleasant-smelling spraint!

middle of the 20th century – a staggering 1,212 otters were killed by otter hounds between 1950 and 1955.

The otter was granted full protection in 1978 after surveys showed that the species had declined catastrophically, particularly in England and Wales, not primarily because of the hunting, but through a combination of habitat loss and the use of organochlorine pesticides which became concentrated to lethal levels in the otter's tissues. Also, in the 1970s, river authorities had become obsessed with controlling water flow by canalising water bodies, which resulted in the removal of numerous den sites and bank-side vegetation. More recently, the emphasis has changed to that of 'soft defence', whereby the surrounding vegetation is used to absorb the high-water levels and control floods.

The highest cause of otter deaths currently is from road accidents or starvation of immature otters, so, while they are thought to have a lifespan of around 15 years, few reach this grand old age. Despite this, from original strongholds in the West Country, Wales, areas of East Anglia, the northwest and northeast of England and Scotland, the otters are clearly reoccupying some of their former territories. While they are still nowhere near as common as they were 50 years ago, this time their future looks much brighter.

Having been persecuted for a long time, this can be a very wary and difficult animal to track down; it's worth it in the end, though.

Roosting rooks

WHEN
October to the end of January

WHERE
Buckenham, Norfolk
(Private)

29 Roosting rooks

The rook is the quintessential farmland bird, and, being a species that directly depends on 'created habitats', historically it has flourished with the ancestral clearance of woodlands as Britain became colonised by man. Looking rather similar to the carrion crow – and being tarred with the same brush by many uninterested in the natural world as such – the rook, unlike its cousin, has a sociable and gregarious personality and a penchant to entertain the committed naturalist with some glorious behaviour that is guaranteed to warm up the coldest of winter days.

Our farmland would be bereft without the sight and sound of the charismatic rook.

Missing only from the treeless uplands of northern and western Scotland and the centres of large towns, the rook has a wide distribution throughout Britain with around 1.1 million breeding pairs, a number bolstered by an influx from northern and eastern Europe to eastern Britain in winter. Rooks flourish in areas of mixed farming with a combination of arable and pastureland in which they are able to forage for their varied diet of invertebrates and grain. It is the consumption of the latter that has not endeared these engaging birds to the arable farmer and has resulted in many being shot as agricultural pests. As they are considered decidedly palatable in some parts, in turn they are themselves eaten as the basis of the ancient dish of rook pie.

Of similar size to the carrion crow, the rook can be identified by a combination of looser plumage around the legs – which make it look like it is wearing baggy pants – and the prominent patch of pale, bare skin around the base of the adult's bill. At a distance rooks can look totally black, but when viewed at close range the plumage takes on a whole range of hues from cobalt to violet and even turquoise, and, when seen in good light, their feathers can seem iridescent. In flight the rook's wings also look comparatively narrow, with their long-fingered primary feathers producing deeper wing-beats than that of the crow. The other key to separating carrion crows and rooks is to look at the company they keep, as the crow tends to prefer a solitary or paired lifestyle as opposed to the rooks' companionable nature, hence the phrase 'a crow in a crowd is a rook, a rook on its own is a crow'.

Although the rook is capable of eating a vast array of food items from beetles to small mammals and carrion to grain, it is thought that, particularly during the breeding season, earthworms are by far the most important component of their diet, comprising anywhere between 60 and 95 per cent. Most of the food is gathered by birds feeding in loose and mobile flocks. This technique exploits patchily distributed but abundant food sources and enables increased vigilance against potential predators which are constantly on the lookout for unwary birds.

The large communal autumn and winter roosts for which rooks justifiably make the top 40 are only part of the story, as most people associate these birds with their shared nesting habits at colonies or rookeries in the canopies of mature trees. These sites become reoccupied so early in the year that the rooks will be present well before the tree buds open, making the numerous nests stand out clearly from the bare branches. Many rookeries have been used for decades and even centuries meaning that the rook can claim the joint accolades of having the most conspicuous nest site of any British bird and being the species that has surely lent its name to the title of more farms than any other animal. Up to 60 species of tree have been recorded as holding rookeries, with conifers often preferred in the windier places such as Cornwall or Cumbria. The classic lowland tree that historically held many colonies was the English elm but, since the large-scale removal of this tree from the British countryside as a result of Dutch elm disease, many rooks have switched to nesting in the tall canopies of oak, ash and beech. Rookeries are not just confined to farmer's fields or along boundaries, but can often be encountered in villages or on the edge of moorland. The size of each colony also varies with geography, as typical English rookeries rarely contain more than 50 nests, and tend to be smaller than their Scottish counterparts, where around 80 nests may be considered the norm.

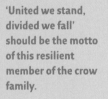

'United we stand, divided we fall' should be the motto of this resilient member of the crow family.

Rooks are monogamous by nature with the pair bond being maintained throughout the year, for a number of breeding seasons and possibly among the long-established and successful pairs for life. Occasionally during the winter and then throughout the early spring, this pair bond is cemented with a combination of pursuit flights, preening each other and courtship feeding. Established pairs are often incredibly faithful to both the colony and even the same nest site, with the male either beginning the nest-building process or repairing last year's effort, before the female joins in to help with construction

and application of the nest lining of moss and leaves. Particularly during the early stages of nest construction, neighbours will frequently steal twigs from each other. This can end in petty squabbles and result in the construction taking anywhere between one and four weeks to be finalised. At most rookeries the older and more dominant birds will monopolise the centre of the colony and push the inexperienced pairs to the margins, but, despite all this seeming discord, when the colony is threatened by birds of prey such as goshawks or buzzards, the rooks unify and mob together to drive intruders away.

Southern England sees the first clutches of three to five eggs being laid in early March, as the rooks are keen to ensure that the young are fledged well before the dry summer months when the invertebrate prey can become much more difficult to find. After an incubation period of 16 to 18 days, both the chicks and brooding female are initially fed by the male, until the chicks are able to thermoregulate, after which both sexes collect food from the surrounding countryside. This is brought back in special expandable pouches at the base of their mouths. Even after the young have fledged at between 32 to 34 days, the adults will continue to feed them for around 6 weeks, after which they will forage together in family flocks until the end of July or early August.

As summer progresses into autumn, the small feeding flocks coalesce as neighbouring rookeries merge to form much larger groups that will abandon the rookery and begin to spend the colder nights at large communal woodland roosts. Like the breeding sites, some ancestral roosts have been used for hundreds of years, with the famous Buckenham roost in Norfolk being mentioned in the Domesday Book of 1086. Claiming up to 80,000 birds in

Nesting begins early in the year, with each pair jealously guarding their nest against their stick-robbing neighbours!

The avian equivalent of the Houses of Parliament, as rooks leave for their winter roost. A flock of rooks is, of course, called a parliament.

December and January makes Buckenham the largest rook roost in Britain and the best seat in the house as dusk approaches and the rooks begin to gather.

The process of roost formation commences slowly in the afternoon, as birds that have been feeding away in the surrounding fields begin to return towards the roost along regular flight paths. These flights are punctuated by stopovers or pre-roost assemblies as the flocks gradually join forces, and then increase massively in size closer to the roost. As the rooks gather on stubble fields adjacent to the roost, from a distance they can resemble a seething carpet of ants as they noisily partake of their last feed of the day. This feeding frenzy gradually subsides until most of the birds are sitting in silence, with some preening and others sleeping; the exception are the fringe birds, which can often be seen attempting to crowd into the centre of the flock. The agitation of these fringe birds also triggers the entry of the flock into the roost as they arise in one great mass and swirl above the wood, resembling the flakes of ash from a huge bonfire.

While in the air the rooks will often perform characteristic towering and tumbling flights called 'crows' weddings': they spiral up to a great height in a tumultuous wall of sound, with some of the birds chasing, swooping and tumbling in the middle of the flock. These aerial displays have often been used to prophesy weather, with high altitude manoeuvres supposedly forecasting fair weather and low flights predicting rain. The reason for this mass flocking, however, is to prevent any potential predator from making a successful kill

both by minimising each individual's chance of being attacked and by relying on the swirling flock to confuse birds such as goshawks and peregrines, which are more effective when chasing down an individual 'locked' target.

After a short while the rooks gradually begin to descend into the wood, sometimes alighting briefly only to settle again, and, within just a few minutes, the vast majority of rooks have landed in the trees, apart from a few late straggling flocks that will fly straight into the roost. Although the birds seem to drop randomly into the roost, research suggests that there does seem to be an element of order, with the adult birds choosing to perch higher in the tree tops and subordinate birds having to settle further down. In severe weather the adults will push the younger birds further down the trees or even displace them entirely. Exactly why the experienced rooks prefer roosting in the more exposed locations is still not fully understood, but birds lower in the roost are certainly at a disadvantage as they are covered with bird guano from their loftier neighbours; it is one of the downsides of communal living.

For centuries it has been considered good luck for a farm to be blessed with a rookery and a bad omen for it to be deserted. Despite the various changes in agriculture, which have seen a number of farmland species drastically reduced, this avian cornerstone of the British countryside continues to thrive. Let us hope that some of the magic from this wonderful bird has rubbed off on the beleaguered farmland on which it and we are so dependent.

Bats

WHEN
Most commonly between April and September, with the mating sites occupied between May and August

WHERE
Found anywhere from Cornwall to the far north of Scotland, including Orkney and the Outer Hebrides

Pipistrelle bats swarming

Like many classes of animal, bats reach their maximum diversity in the hothouses of the tropics, with equatorial countries such as Ecuador possessing around 115 bat species – an incredible 12 per cent of all bats described by science. As Britain is an island on the edge of a continent and at a temperate latitude, it is perhaps no surprise that the number of resident species is a slightly more modest 16. Britain makes up for its lack of species by having the most intensively studied bats in the world, however, and so imagine the surprise when it was revealed in the 1990s that a new species of pipistrelle had been recorded under our very noses!

While pipistrelles mostly consume midges and the like, the more substantial greater horseshoe is capable of a much meatier meal.

The pipistrelle bat, know simply to bat workers as 'the pip', is our smallest and most commonly found bat. Distributed throughout Britain, it is the most widespread species, only absent from some of Scotland's Western Isles. The pipistrelle is also the archetypal bat most commonly seen by people, as they frequently roost in large numbers in houses and habitually feed over gardens and the wider countryside we frequent. While most sightings of bats are made at dusk as the bats are leaving their summer roosts for the night, bats like pipistrelles are actually much easier to see and far more impressive just before dawn as they carry out their swarming behaviour prior to re-entering the roost.

Only very recently have scientists realised that 'the pipistrelle bat' they had been studying for years was actually two species. Researchers noticed that different colonies of pipistrelles were echolocating at distinct frequencies of either 45 kHz or 55 kHz. That the bats were indeed separate species was confirmed by DNA analysis, which suggested that they were much more genetically distinct than was initially thought, and may have diverged into two species at least five to ten million years ago.

Since the split, slight differences became noticeable when both species were in the hand, with the 55 kHz pipistrelle having a much browner face with a more pointed muzzle and a distinctive musky smell, as opposed to the bandit-type black face, more bulbous nose and the less smelly disposition of the 45 kHz pipistrelle. The two species were, therefore, named the soprano (because of its higher frequencies) and common pipistrelle respectively. Alternative names proffered were the 'browns' and 'bandits' because of their different facial colourations.

In addition to the distinct frequencies of their calls and subtly different looks, both species are believed to prefer slightly different feeding habitats: the soprano pipistrelle is more a specialist of riparian (or waterside) habitats and avoids open areas such as farmland, fields and moorland; the common pipistrelle is, however, more of a generalist feeder able to cover a far wider variety of habitats. In flight, though, both species have an identical fast and jerky flight as they pursue, catch and eat an incredible 3,000 small insects a night on the wing. Like many bats, they will often follow linear features such as hedgerows and tree lines between the roost and their favoured feeding areas at a height of between two to ten metres above the ground.

The species are thought not to share roosts, meaning they are reproductively isolated, and, although there is an overlap, the sopranos tend to form larger roosts of over 200, as opposed to the commons, which usually consist of fewer than 100 bats. The common pipistrelles also appear to be more willing and able to change their roosts more regularly the sopranos. Buildings are the favoured roost sites for both pipistrelles, with many being situated in modern houses. The bats prefer to roost in very tight spaces and are frequently on the outside of the building, for example behind hanging tiles, panels or soffit boxes, or under roofing felt or tiles. Each species rarely actually enters roof spaces unless the buildings are very old and have played host to established colonies for many years, such as the famous soprano roost in Lacock Abbey in Wiltshire, which boasts over a thousand emerging pipistrelles at the height of summer. In all roosts the access hole need be no larger than a small slit for the bats to be able to gain access to the roost unimpeded.

The mating colonies of both pipistrelles are occupied between May and August (or occasionally September) and consist almost exclusively of female bats. These females give birth to their single young (very occasionally twins) between early June and the middle of July; for the first three weeks the young are fed solely on their mother's milk. In between feeds, the mother will make at least two feeding trips, as opposed to the one long feeding trip made by the female before she gave birth and while she was pregnant earlier in the summer. By the time the youngsters are three weeks old they have tripled in weight – to a princely three grams – and are able to make their maiden flights. By six weeks, they are able to forage for themselves, and they reach their final weight of four grams; they are ready to be weaned after fifty to sixty days.

'Bandit' or 'brown'? Difficult to say which pipstrelle, until it's heard echolocating.

Both species have social calls in the roosts; these are at around 20–30 kHz, much lower-pitched than their echolocating calls, and are often easily heard by children and some adults. Both species can be very active in the roost during the day, especially when the young are present or the roost gets hot, and they can become particularly noisy prior to emergence. Pipistrelles usually begin to leave their roost around 20 minutes after sunset, unless the weather is particularly inclement, with strong winds, rain or particularly low temperatures, when they will usually leave later. Interestingly, they do not all bundle out at once and in a haphazard fashion; the adult females tend to come out first, followed by the young bats that are large enough to fly. After the first bat has emerged, there seems a delay before the rest follow, and they often come out in small groups of three or four rather than a steady stream. It may be that some pipistrelles sit at

Like the soprano pipistrelle, the daubenton is a water specialist, with a palate that prefers aquatic insects.

the entrance, blocking it up and causing a small logjam as other bats press up behind and urge them out. At large pipistrelle colonies it can often be 45 to 60 minutes from the emergence of the first to the last bat as the entire colony leaves for a night's dining.

Upon emergence, the bats will usually follow the same flight path every night as they travel to their favoured feeding areas, and the diet of the soprano pipistrelles is thought to consist predominantly of insects that have an aquatic larval stage, such as flies and midges, in keeping with their preference for freshwater sites. The vast majority of insects are caught by hawking and then eaten on the wing, rather than gleaning prey from vegetation like the long-eared bats.

After having spent the night filling their stomachs on flying insects, both species will return to the roost before dawn and congregate in large numbers, while circling erratically around the roost entrance in a spectacular process known as 'swarming behaviour'. The best locations for watching this performance are outside large soprano roosts as far more bats will be whizzing around. The exact reasons for this behaviour are not thoroughly understood, but it may either be a social exercise to make sure the colony stays together and all the bats are present and correct before entering the roost, or it may be to make a visual check of the roost to make sure it is safe to enter. This flying around is often accompanied by some 'chatter', so there may well also be an exchange of information – on good feeding areas that various individuals have located – before they re-enter the roost and settle down for the day. This swarming behaviour is not just confined to the two species of pipistrelles, as tree-roosting bats such as barbastelles and noctules will often fly around their tree roosts before dawn. This spectacle is often used by bat researchers to locate the roosts of certain species, as it is much easier to locate large swarms of bats at dawn than when they are leaving the roost in dribs and drabs at dusk.

Mating of both species occurs in the autumn, between September and November, at well-established mating roosts, which are distinct from the maternity roosts. The males may possibly hold territories and certainly seem keen to mate indiscriminately with any female bat of the same species. Fertilisation does not occur immediately, however; it is delayed until after hibernation, as the female stores the live sperm inside her body until she is ready to ovulate the following spring, a reproductive feature that is unique to bats. Few pipistrelles are ever found in winter as they are thought to hibernate either singly or in small groups in cracks and crevices of buildings or trees. By cooling their body temperature down to that of their surroundings, which in turn causes an extreme drop in heartbeat rate, breathing and other bodily functions, the pipistrelles seem to be remarkably tolerant of cold weather. They will also wake up periodically to defecate, drink and will even take advantage of any warm days in which to feed.

Both species have probably declined as a result of modern agricultural practices, such as the use of pesticides, and the renovation of roosts from which the bats may have been excluded, or to which they are prevented from returning because of the use of toxic chemicals used to treat the timber. However, on a positive note, with the recent trend in organic produce, organic farms have been found to play host to higher concentrations of insects and bats, which is certainly good news for our daring duo of pipistrelles.

Butterflies

WHEN
Mid-June to mid-August for silver-studded blues; July and August for purple emperors

WHERE
Great Orme, North Wales and Portland, Dorset for silver-studded blues. Bentley Wood, Hampshire/Wiltshire and Bookham Common (National Trust), Surrey, for purple emperors

27 Clouds of butterflies

Butterflies are beautiful, iconic creatures that undoubtedly enrich our lives. Not only do they act as emblematic flagship species, they are also sensitive indicators of both the state of our countryside and the condition of some of our most precious wildlife habitats. Butterflies in Britain, however, are in trouble; with over three-quarters of our 59 breeding species in decline and 5 species extinct, they have undoubtedly fared worse in the last 25 years than either birds or plants. As the numbers of butterflies have plummeted, so too have our chances of seeing clouds of butterflies, which many an older naturalist has clear memories of as a child. But, given a good butterfly-watching day at the right location and with a degree of the requisite luck that goes with any wildlife expedition, a few species can still be seen in very healthy numbers.

The silver-studded blue, it has to be said, is at best an unremarkable little butterfly. To borrow from birdwatching parlance, this butterfly is the entomological equivalent of a 'little brown job', or, in the case of the species itself, a 'little blue job'. As a species, it is a butterfly of essentially two habitats: of lowland heath and coastal limestone in a few special places. Like many butterflies, the silver-studded blue has disappeared from many previous sites and declined drastically in others, and, while largely confined to the heathlands of Cornwall, Hampshire and Surrey, it is still particularly abundant on the Isle of Portland in Dorset and the Great Orme in North Wales, two huge lumps of limestone jutting out into the sea but still attached to the British mainland. While many of the surviving colonies may not contain more than a thousand adults, some sites, such as those found on the Great Orme headland, still have the capacity to produce huge numbers of butterflies, and single colonies have been recorded as containing over 30,000 adults!

With a wingspan of 29 to 31 millimetres – not much larger than a thumbnail – the males are coloured a deep blue on their upper sides, with white fringes around the edges of their wings that contrast sharply with wide black borders. The females, however, are not as splendidly coloured, having to settle for a brownish upper side, which is usually tinged blue near the body. The butterfly derives its name from distinctive metallic black spots with silvery-blue centres along the outer edge of each hind wing. While generally looking smaller and a much deeper blue than the common blue butterfly, a species with which they will commonly associate, they confusingly exist in a number of colour variations. The silver-studded blue populations on the Great Orme, for example, are considered even smaller and often emerge three weeks earlier than their heathland cousins, differences deemed large enough to raise the status of the Great Orme butterflies to that of a different sub-species.

Any early spring walk would be enhanced by the colourful splash of a male orange tip, the precursor of some 'clouds' later in the season.

Like most of the British blue butterflies, silver-studded blues live in close-knit and discrete colonies from which the adults will rarely stray; most butterflies will move less than 20 metres per day and only a few will travel more than 50 metres. Although they will readily take to the wing on sunny days, their flight is slow and fluttering and often only takes place a few centimetres above the ground before they settle once again. The lifespan of the adults is rarely more than four or five days, in which they must mate before they die a spent force. The adults' very poor powers of dispersal means that colonies separated by only a few hundred metres are effectively isolated and therefore vulnerable to local extinction if the habitat becomes degraded or a catastrophe besets a population. So, with its poor recolonisation powers, the silver-studded blue is not well equipped to cope with modern Britain's fragmented habitat.

However, where habitat and climatic conditions are favourable, this butterfly can often be incredibly abundant with populations in their tens of thousands. In late afternoon the butterflies will congregate together to roost communally on grass tussocks or sheltered bushes; the best time to see large numbers on the wing is early in the morning when hundreds can be seen stretching their wings as they prepare to take their first flight of the day. With such a short lifespan, they are busy: the males rarely find the time to visit flowers to feed as they constantly patrol backwards and forwards in their constant search for virgin females that have just emerged from their chrysalises.

Once the male locates a suitable female, the courtship consists of a brief flit together in among the ground vegetation before the female drops to the ground.

What the silver-studded blue lacks in beauty, it makes up for by sheer numbers, with locustlike swarms at some favoured sites.

The female is incredibly fussy as to where she lays her tiny white and disc-shaped eggs, with each one being placed singly in the mats of bird's-foot trefoil or common rockrose in the limestone colonies, or low down on young sprouting gorse or heather shoots on the southern heathland sites.

Like many of the blue butterflies, the silver-studded blue has an intimate and highly evolved relationship with one or two species of black ant from the genus *Lasius*, on whose survival the butterflies seem to be entirely dependent. Research has recently shown that the females will choose to lay their eggs where they are able to detect the ant pheromones so, when the caterpillars emerge, they are able to link up with the ants quickly. On hatching in March, the tiny caterpillars feed on the young, tender shoots of bird's-foot trefoil and common rockrose or gorse and the various heathers close to where the eggs were laid. During this whole time the growing caterpillar is constantly tended to and protected by the ants, which receive a sugary solution secreted from a slit called the honey gland in the back of the caterpillar in return for their devotion. The caterpillars also possess a pair of small tentacles at the rear of their bodies, which can be erected to stimulate the ants into 'milking' the caterpillar.

By the end of May to the middle of June the caterpillar is ready to pupate; the relationship between the two insects does not end there, however, as many of the caterpillars will pupate in, or close to, the safety of the brood chambers of the ants' nests where they continue to produce secretions for the ants until they are ready to emerge. On leaving the ants' nest, the fresh adult butterfly must first inflate its wings before it is able to fly, which is a vulnerable time as the butterfly could easily be taken by predators at this stage. However, by surrounding the silver-studded blue, the faithful ants may well help to protect the butterfly from attack until it is ready to fly.

A pair of green-veined white pausing for a rest on a cuckooflower – green-veined being a rather uninspiring, if accurate, name for this delicate butterfly.

It is difficult to imagine a butterfly that contrasts more with the silver-studded blue in terms of looks, habit and habitat than the purple emperor. The 'emperor' is considered the Rolls-Royce of butterflies with its large size and stunning looks, as it soars, wheels and glides majestically above the tree canopy of the southern woodlands from Hampshire to Wiltshire and across to Sussex and Kent. One thing that these two very different butterflies do have in common, though, is their willingness to congregate (albeit with very different numbers) for the sole purpose of mating.

The purple emperor has been the subject of much admiration over the centuries, with the English poet and naturalist, the Reverend George Crabbe, writing of the species at the turn of the 19th century, 'above the sovereign oak, a sovereign skims, the purple emp'ror, strong in wing and limb'. It is the males of this large butterfly that take the plaudits, and, with the iridescent purple sheen of their upper wings, they can be confused with no other species. While the males will occasionally be tempted to descend to the ground to probe for salts from the road surface or animal dung, this elusive butterfly is largely confined to the tree tops where they feed on aphid honeydew and tree sap.

Adult purple emperors emerge from the last week of June onwards, and can be seen on the wing until at least mid-August. Although a population may well hatch at very low density in a particular wood or a complex of smaller woods, its highly mobile nature means the sexes will soon meet as they tend to fly to the highest point, where the males congregate to establish territories around special 'master trees'.

The purple emperor is the entomological holy grail of our southern oak woods. A glimpse of the male's iridescent wings is a never-forgotten experience.

It's confined to the Norfolk broads, so your best chance of a sight of Britain's largest butterfly, the swallowtail, is to stake out their nectar plants such as ragged robin or thistles.

Male emperors take to the air at around 11 a.m., initially partaking in comparatively low-level flight as they soar through the upper branches where huge distances can be covered, often moving from one wood to the next. By noon they have moved to the master tree, a prominent tree – this is often, but by no means always, an oak and is frequently used year after year – where the males establish territories high up on outer south-facing branches. The males will then perch with their wings half open and facing outwards so they are in a position to launch themselves at other males, leading to towering battles as the combatants rise into the sky above the canopy, with their wings clashing and flashing purple and white as they go.

Each male will occupy the same perch all afternoon, although he may be interrupted by plenty of flights to take on other males; the female is much more elusive, as she lacks the iridescent purple sheen and is rarely seen before egg laying. When ready to mate, she moves to the master tree where she will be guaranteed to receive immediate attention from a number of suitors. After mating, the female disappears off into the woods to track down mostly goat willows on the edge of a woodland ride or glade, on which she can lay up to a dozen eggs, one at a time, on the flat leaves. The small green eggs hatch after about ten days, and the larvae then move to the tip of the willow leaf and start to feed. In autumn the well-camouflaged caterpillars will hibernate in a pad of silk until the next spring when they are ready to resume feeding on the leaves until pupation in early June. Only after they emerge from the chrysalis and the blood has surged into their wings are they finally ready for a short but glorious life spent in the canopy.

Terns

WHEN
May to mid-August

WHERE
Farne Islands, off Seahouses, Northumberland; Cemlyn Bay Nature Reserve (North Wales Wildlife Trust); Orkney & Shetland Isles

26 Terns attacking

On a summer's day at a few special coastal or island locations around Britain, such as the Farne Islands off the coast of Northumberland or Cemlyn Bay on Anglesey, there are a family of birds that just cannot be avoided. Visiting our British shores for a few short months to take advantage of the seasonal abundance of seafood, the terns or 'sea-swallows' are both enchanting to watch and feisty in character.

When seen flying above, it is not hard to understand why terns are also named sea swallows.

Of the five species of tern that regularly come to breed along Britain's coast, only the little tern seems to systematically prefer its own company; at the premium tern colonies it is not unusual to see Arctic, common, Sandwich and occasionally the rare roseate tern all nesting either in among, or alongside, each other. Unlike many coastal birds, such as guillemots, razorbills and kittiwakes, which nest in large vertical populations in order to prevent predation, or Manx shearwaters and puffins which seek refuge underground, terns are seemingly much more vulnerable to attack as they breed in among the flat, open and accessible coastal vegetation. Of the four communal tern species, the Arctic tern is by far the most aggressive in defence of its eggs and chicks. Usually with some help from the common tern, it will often tackle the issue by dive-bombing and strafing with bird guano any potential terrestrial predator – such as foxes, stoats, marauding gulls or even humans – which strays too close to the nest.

Despite being the most abundant breeding tern in Britain, with a current population of 53,000 pairs, the Arctic tern is probably not the most commonly encountered species because their largest colonies are confined to the remote, northern outposts of the Orkney and Shetland Islands. This contrasts with the less abundant, but paradoxically entitled, common tern, which breeds both on the coast and inland at freshwater sites closer to the large population centres in southern Britain, and hence is more frequently seen.

Throughout the ages, common and Arctic terns have frequently been misidentified and confused, with the result that even experienced birders will often note 'commic terns' in their notebooks, a catch-all term used to describe both species, or where the identity of an individual has not been fully secured. There are, however, numerous subtle differences between these two closely related species, with the Arctic tern being darker, smaller, more narrow winged, proportionately longer tailed, shorter legged and with a crimson red bill in contrast to the black-tipped, orange-red bill of the common tern. In flight, the Arctic tern seems more buoyant and even more elegant than its 'not so common' cousin, and the flight feathers of the Arctic also appear more translucent when seen from below.

The British shores and offshore islands mark the very southern breeding range of the Arctic tern; many birds will have wintered off the Antarctic pack ice or even come from as far away as Australia to breed. This annual migration, which, for birds breeding further north into the Arctic Circle, will complete an annual round trip of 21,000 miles, make the Arctic tern the avian world's most famous global traveller, and the only species that regularly visits all seven continents – an impressive feat for a bird weighing less than 120 grams.

The Arctic tern is also a bird that enjoys the company of its own species; with the possible exception of when it is feeding, the tern will spend its whole year breeding, roosting and migrating in gregarious and noisy flocks. As befits many seabirds that have a surprising longevity, the Arctic tern is also a bird that practises the art of monogamy, only usually choosing to find a new partner on the death of a mate. Being faithful is rewarded by breeding success, with older and more experienced pairs systematically rearing more fledglings than a younger, inexperienced pair. While Arctic terns will occasionally try and reproduce at two or three, the vast majority of pairs will not breed at their natal colony until they are at least four years old, with established pairs over the age of eight having the highest fledging rate of all. These older terns tend also to be the first to return to the breeding site, with the more inexperienced birds often not arriving until three weeks after their elders. It is thought that the pairs do not spend the winter together and will either convene on the way back up north or at the pre-breeding roosts close to the colony itself.

Doesn't look very threatening, does he? Well that depends if you are about to step on his egg or chick, as the Arctic tern is not backwards in coming forwards and has a very sharp beak!

Arctic or common? With the black tip, it can only be common, which, ironically, is the less common of the two species!

Initially the terns will spend very little time on land, only visiting briefly at dawn before spending the rest of the day feeding out at sea and roosting away from the colony on the beach or sandbanks close by. Eventually, however, the birds will start to spend more time at the nest site, and the colony steadily fills up. Each mature pair shows immense fidelity to particular nest sites and will frequently breed on the same small patch of ground year after year if it has led to success in previous breeding seasons. As the most experienced birds position themselves in the centre, the younger couples are frequently forced to the edges of the colony.

During the short pre-breeding period, each pair will cement their bonds by a highly ritualised aerial courtship which involves them chasing each other quickly into the air before rapidly descending. The male seals the deal on the ground as he walks around his mate and woos her with a pre-caught fish at their nest site. The nest site itself is often nothing more than an unlined scrape in the ground and, at best, will be adorned with a few random pieces of debris or include one, or a number of, coastal plants.

The British colonies are considered the largest and also the most densely packed of any of the Arctic tern nesting locations, but, even there, the nesting density is less than that of the common terns, which may well be in keeping with the more aggressive nature of the Arctic tern. The nests on the Farne Islands are around three metres apart; this circular territory is vigorously defended against encroachment from neighbours and will be the entire world for the Arctic tern chicks until they fledge.

Either one or two eggs are laid by the female and both sexes will take it in turns to incubate the clutch while the other bird is free to feed out at sea. The eggs will hatch at around 22 to 24 days old with the chicks being born both precocial, meaning they are relatively mature and mobile from birth, and nidifugous, which describes the chicks' ability to leave the nest early. After a short period of two or three days being brooded and fed, the chicks will move around their territory to hide in among the stones or vegetation either at the first sign of danger or for shelter from the elements.

Adult Arctic terns will feed either solitarily or in small flocks, or occasionally in very large flocks if a large shoal of sand eels has temporarily created an abundant bonanza. In contrast to the other tern species, the Arctic has a 'stepped-hover' habit when hunting, which involves often hovering up to ten metres above the water surface and descending vertically to a new position and again hovering before either plunge-diving, descending further to an even lower position, or breaking off the hunt to search elsewhere. This technique of hovering at a range of different heights in a single vertical plane is very characteristic and can be used to identify the Arctic tern from considerable distance at sea.

Terns' main prey consists of sand eels, sprats, juvenile herring and capelin; it is rare for the adult terns to bring back more than one fish at a time to their waiting chicks. The chicks recognise the advertising call of their returning parent, and will come out of hiding on hearing the call to be fed on average around twice every hour, with the fish being both proffered and swallowed head first to ensure the gills or any spines do not lodge in the chicks' throats.

As anyone who has ever walked among an Arctic tern colony will attest, it can be both an exhilarating and scary experience thanks to the confrontational way in which the adults will attack any animate object they perceive to be a threat. As a gull or crow approaches a nest, the adults will give an alarm call and then rise above the threat to chase it away, and against mammals such as foxes or stoats, many birds will often work cooperatively to mob and drive the potential predator away from the colony. The terns can be particularly intimidating against any humans blundering into the colony, as a small flock rises over the perceived threat and individual birds take it in turns to swoop down while uttering an angry 'kek-kek-kek-kek'. It is not uncommon for diving birds to draw blood with their daggerlike bills during these strikes, and at the very least the intruder may 'be limed', as the birds let us know what they think of the interruption. Solitary nesters tend to be the fiercest defenders of their nest's contents, with terns near human settlements generally considered to be the most aggressive of all. Unfortunately, despite this vigorous defence of their nests, a determined predator can, and often will, cause considerable damage, which is why most tern colonies are located on offshore islands.

Chicks surviving the attentions of local predators and which have been well fed will fledge after 21 to 24 days, but the parents' bond with their young is

When an Arctic tern is this close, be prepare to be pecked!

thought to be maintained for one to two months after they leave the colony and for at least part of the immense journey back down south. Unfortunately a 30 per cent decline in the number of breeding Arctic terns has been noted in the last 10 years, which may well be down to a combination of overfishing of sand eels and warmer sea temperatures, which push the fish into deeper waters and out of the reach of the diving adults. This provides us all with more reason to treat the colonies with respect as we get 'up close and personal' with surely our most graceful and pugnacious seabird.

25 Pine martens hunting

In 1921 the author J Fairfax-Blakeborough wrote of the pine marten, 'he is the wildest of the wild things left to us, and all his habits and instincts make for secrecy and isolation'. After centuries of persecution, these habits and instincts mean that a once widespread and wonderfully engaging member of the weasel family is still so rare and elusive that no one is exactly sure if it still survives in England and Wales. But in the Scottish wilds, with knowledge of their favourite stamping grounds and of their terrible fondness for sweet things, they can be tracked down surprisingly easily.

The pine marten arrived in Britain having spread northwards after the last glaciation, before the English Channel had risen sufficiently to breach the land bridge with the Continent. Pine martens were once spread right across woodland Britain, with abundant records from Sussex, Devon and Cornwall, but, as a result of the mass clearance of our forests, trapping for fur and intense persecution, they were either very rare or extinct in many lowland areas of Britain by 1800. The popularity of shooting and concomitant rise in the number of gamekeepers in the mid-to-late nineteenth century led to their continued disappearance in all but the most remote areas. Outside tiny relict populations in northern England and Wales, which may currently be on the verge of extinction, it is now virtually impossible to see a pine marten in the wild outside their Scottish Highland stronghold. While the Caledonian pine forests and the rocky, mountainous highlands with their remnant woodland are now considered the core haunts of the marten, they are nowhere near as suitable for the animal as the species-rich deciduous lowland woodlands, which are considered prime habitat. Their current distribution is merely representative of the areas where persecution has historically been at its lowest.

The pine marten is a very handsome beast with an alert and inquisitive character to back up his good looks. While appearing about the same size as a domestic cat, the pine marten has a much more slender and lithe body and a long bushy tail, giving it a more foxy, than feline, appearance. Pine martens also look different in their shorter and darker or 'plain chocolate' summer coat as opposed to the much denser greyish-brown or 'milk chocolate' winter coat. In addition to a pointed snout and small eyes, they have a pair of quite large creamy-coloured ears, which stand out from their brown fur, but perfectly match the creamy-yellow throat and bib that can extend down to their forelegs. This pale bib can differ subtly in colour but will also vary in size and pattern making it a valuable tool to enable identification of particular individuals in the field. Male martens are generally about a third larger than their female counterparts, and all the adults are incredibly agile climbers with strong limbs,

Hounded to extinction in vast areas of the countryside, the pine marten sought refuge in the Scottish Caledonian pine forest, from which the species is now slowly recolonising old stomping grounds.

Pine martens

WHEN
Can be seen around dusk in
any month

WHERE
Scottish Caledonian forest
and pine plantations.
Rothiemurchus Estate, near
Aviemore, Inverness; Beinn
Eighe NNR (Scottish Natural
Heritage), Highlands

large feet and partly retractable claws, which enable them to climb the smoothest trunk like a squirrel on steroids.

While displaying a phenomenal agility in trees, they are equally happy on the ground, where much of their time is spent foraging as they investigate feeding possibilities with a distinctive long-legged loping or bounding style. As in the case of otters, since pine martens may be sparsely distributed in some areas, scent plays a very important part in conveying messages to other individuals. Their faeces or scat is coated with an individual odour from their anal glands, and pine martens will also anoint objects with their scent; this is particularly important during the summer when they are keen to pass messages to either the same sex to keep away, or to members of opposite sex to come hither. Although martens are predominantly solitary and silent creatures, they do have a wide range of grunts, squeals and snarling noises which they will make when they meet other martens, or females will use when they need to communicate with their cubs.

As with most British terrestrial mammals, sightings away from a few favoured spots are few and far between so, when looking for the presence of pine martens, the mammalogist is heavily reliant on looking for tracks and signs. Unlike the much heavier otter, pine marten tracks can be very difficult to spot as they tread very lightly; they also spend quite a lot of time off the ground. Tracks with five toes and a four-lobed pad can, however, sometimes be spotted in snow. It is the actual distribution of prints that can often be more revealing, as, when moving at speed, their bounding style of movement means that quite often their paw marks can be found in small groups of four with stride lengths of anywhere between 50 centimetres and 90 centimetres between each cluster.

Pine marten scats differ enormously in size, shape and colour, mainly because of their incredibly catholic tastes which can vary from red deer carrion to rowan berries, but they are usually long and cylindrical and often shaped inthe form of a U or &. Pine martens deposit their scats throughout their home range with the den sites and paths being often well marked spots, but in areas where populations are very thinly scattered, scats may be few and far between. Unlike otter spraints, which have a very distinctive and surprisingly pleasant smell, it is difficult to separate marten scats from those of a fox, unless of course the scat is up a tree, in which case it will definitely be the former!

Their dens can vary enormously in nature, but always tend to be placed above ground so as to avoid the unwanted attentions of natural predators like the fox. Pine martens will often favour cavities like tree holes, which have either formed naturally or have been pre-prepared by woodpeckers. In the absence of suitable tree holes, such as in a plantation or where the trees are very young, the martens will equally use rock crevices, squirrel dreys, barns, or even the occasional attic of a remote cottage. Pine marten boxes have recently been placed in many nature reserves with much success, and the added bonus is that occupied boxes are easy to distinguish from the territorial pile of scat placed on top!

As with most mammals, census work involves more looking for its faeces than the animal itself. The producer of this pile seems to have feasted on rowan berries.

Peanut butter, jam and blancmange tend to be their favourites and, if these substances are smeared on the trees or branches, the pine martens can also be induced to show their arboreal agility while they satiate their sweet cravings.

Despite being primarily nocturnal and decidedly less active during the winter, pine martens are much more commonly seen during the long summer days, with evenings generally the prime time to try and catch a glimpse of them, particularly the females who will be pressed to gather enough food to sustain their litters. Rather than actively searching for pine martens in suitable habitat, it is much better to stake out favoured feeding spots such as picnic sites, bird-feeding stations or hides. A number of enterprising reserves regularly feed pine martens in front of both a captive and captivated audience, and, at such sites, those pine martens that have overcome their natural persecution complex can be seen regularly feeding on a variety of food stuffs.

While the pine marten in Europe is a species that undoubtedly prefers extensive areas of forest, it has hung on in Britain despite the highly fragmented nature of our woodland. In areas with only a remnant population of trees, pine martens have been able to survive in rocky habitats which offer them a form of the three-dimensional topography otherwise only found in forests. Even the seemingly sterile belts of conifer plantations have been used by martens, providing there is a rich supply of short-tailed field voles.

Pine martens are omnivorous by nature with an incredibly varied diet including carrion, small mammals, birds, amphibians, reptiles, eggs, beetles, earthworms, caterpillars, hazelnuts and a whole variety of autumnal fruits and berries depending on their seasonal abundance. Despite opportunistically eating anything that comes their way, Scottish pine martens mainly consist on beetles and field voles in the spring and summer, carrion in the winter, and nuts and berries in the autumn, with one road-kill marten found to contain over 300 rowan berries in its stomach.

The male pine marten will attempt to exclude adults of the same sex from its home range, while overlapping with the smaller territories of one or more females. The size of a male's home range in Scotland may vary between one square mile to thirteen square miles, and tends to vary according to the quality of habitat, with pine martens living much closer together in the much sought-after species-rich woodlands.

Mating takes place in July to August when the female comes into full oestrus; after some playful chasing by the pair, amid much squealing, the male takes control of the situation by biting the neck of the female. This is the sum total of the promiscuous male's involvement, as he will then proceed to mate with a number of other females and plays no part in rearing any of the litters.

A single litter is usually born in March or April after a delayed implantation of around six months and a gestation period of thirty days; litter sizes usually vary between two and four. Born blind and hairless, the cubs do not open their eyes for forty days and spend the first seven weeks in the den before emerging to take their first steps and explore their mother's home range. By three months they look like miniature adults and, while very playful, still need time to practise climbing before they finally disperse to find their own home range any time between late August and the beginning of October

Trapping, shooting and snaring are still thought to account for half the pine marten premature deaths even though they have been fully protected since 1988; busy roads also take a heavy toll, in addition to those individuals picked off naturally by foxes or the occasional golden eagle. While widely, and unfairly, demonised for taking game from traditional pens, many farmers and estate managers still see the pine marten as a pest; they may, however, now be severely out-numbered by the massed ranks of tourists and naturalists keen to see one of our most enchanting and resilient mammals. Since its statutory protection and the decline of gamekeeping, the pine marten is slowly expanding from its core range, and isolated populations may also be aided by translocations so this wildest of animals will hopefully remain truly wild.

Thanks to their sweet tooth, the ever-curious pine marten can be surprisingly straightforward to catch up with at a few key locations

24 Roosting parakeets

England's capital city has, over the last two decades, played host to one of Britain's most unlikely and exotic avian spectacles, the appearance of the ring-necked or rose-ringed parakeet. This colourful addition to our avifauna, whose introduction or accidental release has been the subject of many stories and rumours, has found suburban London to its liking and, with a mushrooming population, for good or bad, the parakeet is definitely here to stay. While most Londoners living south of the Thames will have become familiar with the hoarse shrieking call of these little green arrows as they take advantage of free garden handouts, fewer will have seen the spectacular large autumnal and winter roosts.

The parakeet itself is about the size of a collared dove, with a very long pointed tail, a rounded head and an overriding pale-green plumage. Both sexes have a deeply hooked crimson-red coloured bill, where the upper mandible is specially unhinged so they can feed on fruit, and the male has a delicate black and pink ring encircling his face and a bluish hue to his nape during the breeding season. The parakeets' flight silhouette is distinguished by their pointed wings, long tail and flickering wing-beat as they charge around at high speed. They are also frequently heard well before they are seen as flights are accompanied by their loud raucous screeching 'kee-ak' call. While totally at home in the air or perching, they are not at all comfortable on the ground as their short legs enable them to do little more than waddle around.

There are four sub-species of rose-ringed parakeet – two from Africa and two from Asia; research work indicates that the British parakeets seem most closely related to the Indian race. The rose-ringed parakeet has the most northerly natural range of any parrot in the world, with some of the Indian parakeets encountered in the Himalayan foothills, and this may explain why they are so well equipped to survive in a cold northern climate such as ours.

The rose-ringed parakeet has a long history as a cagebird both in Britain and Europe, and has been a popular British pet since Victorian times. It is thought that a combination of accidental escapes and deliberate releases, principally around the 1960s, formed the basis of the present population, which is about 10,000 birds around Greater London and may be double that figure nationally. It is impossible to mistake the rose-ringed parakeet for any of our native birdlife in Britain, although, confusingly, introduced or escaped monk, alexandrine and blue-crowned parakeets are also thought to be present in small numbers and can occasionally be picked out in the large parakeet winter roosts.

While the first documented breeding of rose-ringed parakeets was reported in 1855 in Norfolk, wild birds were regularly seen around London in the 1960s, with the first documented breeding record from the capital being in 1971. Since

This charming if noisy invader from the Indian sub-continent has got its beak in the door and is now definitely here to stay.

Parakeets

WHEN
October to January for the
best roost numbers

WHERE
Esher Rugby Club; Hither
Green Cemetery

The parakeets seek safety in numbers during the winter as they roost in ever larger numbers, having gorged themselves on Londoners' handouts during the day.

then the population has steadily increased, reaching an estimated 1,000 by 1986, leading to its addition to Category C (for introduced species that are deemed to maintain a self-supporting population) of the British list of birds in 1984. By 1998, population estimates made from counts at their winter roosts put the number of parakeets closer to 2,000, and the main roost at the Esher rugby club has recently reached close to 6,000 birds. In addition to the parakeet's undoubted stronghold in southeast and southwest London, the species has been recorded in virtually every county in England and has even reached Wales and the Scottish borders.

The parakeet's rapid expansion is due partly to Britain's passion for feeding garden birds; the parakeets will happily come down to feed on peanuts and sunflower seeds, which provide the mainstay of their diet during the winter when more natural foodstuffs such as buds, fruit, seeds and grain are obviously in much shorter supply. The main feeding times for the parakeets seem to be early morning and late afternoon, leaving the rest of the day outside the breeding season for the birds to preen or loaf around. While mostly welcome in gardens, they are less so in some of the orchards in Kent, which are beginning to see the first signs of economic damage from marauding flocks. This mirrors behaviour of the parakeets in their native India and Africa where they are considered a serious pest when enormous swarms raid ripening crops.

As feeding parties in most gardens are generally confined to small groups, the one place to see large numbers of parakeets is at one of their winter night-time roosts. The five main roosts are currently at Esher rugby club, Hither Green Cemetery, Reigate, Ramsgate and Maidenhead. The numbers at these roosts really begin to increase in the autumn once the young have fledged and the adults have moulted, which leads in turn to the formation of large flocks. As dusk approaches, large numbers of birds begin flying in from the surrounding countryside from a distance of up to 15 miles away, and squadrons of these aerobatic and manoeuvrable little birds can put on quite an aerial display before tumbling out of the sky to jostle for a place in the tops of the trees of the roost. The accompanying soundtrack, as the birds fly in from their feeding areas before settling down for the night, can also be deafening.

This communal roosting behaviour is partly thought to be a mechanism for deterring potential predators, such as sparrowhawks or goshawks, as the parakeets achieve a higher level of personal safety by sticking together, and also any attackers will invariably find it harder to focus and single out a parakeet from among the rapidly moving flocks. The other purpose of these night-time roosts is so that an information exchange can take place on the location of good foraging areas, with birds that fed poorly the previous day following well-fed birds out of the roost the following morning.

British-breeding rose-ringed parakeets may nest as early as January, but it is thought that most egg laying does not take place until early March. The favoured nest sites are always treeholes and, while they are capable of enlarging a hole that has already been chiselled out, they are largely reliant on taking over an old woodpecker's hole or exploiting a crack that has formed naturally in a mature tree. This hole-nesting behaviour may lead to competition with native species such as starlings, woodpeckers, owls and jackdaws where suitable holes are in limited supply and no nest boxes have been erected. The parakeets will also nest either solitarily or in a semi-colonial fashion, with several nests being found in one tree, or a stand of trees, where opportunities prevail.

The female takes sole charge at incubation and the eggs hatch at around 22 to 24 days, after which the chicks are brooded and fed by both parents. While the clutch size is comparable to that of their native India, from limited research work carried out, the breeding success in Britain is much lower, with 0.8 being the average number of fledged youngsters per nest. This is thought to be because the parakeet is still acclimatising to a new climate and diet, and also that fat-rich peanuts that are collected by the adults from garden feeders are no match for the varied and natural diet that chicks would be fed back in India.

In addition to the threat of starvation, other possible dangers to the chicks are grey squirrels and the occasional removal by humans as free pets. The chicks that survive will not leave the nest until at least 40 to 50 days, after which they will remain with their parents for a few weeks before branching out on their own. They will take around three further years to mature and breed.

In captivity, rose-ringed parakeets have lived for over 20 years, so it is likely that, if they successfully fledge and gain a familiarity with all the best feeding areas, they may well become as common a component of London's wildlife as the house sparrow once was. The populations of parakeets in India have declined drastically as they are caught in large numbers for the pet trade; in Britain, numbers are set to increase and amazing figures of around 50,000 have been estimated by 2010. It would be ironic if British-born birds of this colourful and engaging species were sent back to India to bolster the native population from which they once originated.

23 Dancing cranes

Unlike the fanfare that accompanied the successful reintroduction projects of the white-tailed eagle, red kite and osprey, the Eurasian crane's completely natural recolonisation of the Norfolk Broads, and now the Fenland, has been largely unheralded. While the crane's position as a breeding bird after a gap of 400 years is still tenuous, the fact that our tallest breeding bird is back in Britain at all is testament to sterling conservation work and large-scale habitat restoration.

The crane's initial disappearance from Britain in the 16th century was thought to be largely as a result of the draining of eastern England's once extensive fenland, which spread across the counties of Cambridgeshire, Lincolnshire, Norfolk and Suffolk. This mass drainage may well also have resulted in the eventual disappearance of species such as the purple heron and spoonbill, which can still be seen in the Netherlands. The crane was rare even in the day of Henry VIII: an Act of Parliament banned the collection of crane eggs, and the last reliable British breeding record is known to have been in 1542. Hunting must also have played a part in the bird's demise but a feast to honour Elizabeth I's visit to the Fenland less than 20 years later, though it included an astonishing 70 bitterns, 28 herons and 12 spoonbills, had just 1 crane.

Despite the crane's current extreme rarity in Britain, the breeding range of the species extends from north and west Europe, right across Asia to northern Mongolia, north China and eastern Siberia, with its wintering sites traversing a broad band from Spain to East Africa, the Middle East, India and southern China. The cranes flourish in huge, isolated wetlands but have also adapted to smaller more disturbed wetlands within intensively cultivated landscapes, such as those now found in the Norfolk Broads. Unlike the small flock of cranes in the Broads which are resident, the large breeding populations in Sweden, Finland and Poland are migratory and are known to gather in huge flocks en route to their wintering grounds at staging areas such as Hortobágy National Park in Hungary or Lac du Der-Chantecoq in France, which can play host to over 50,000 and 40,000 cranes respectively each autumn. The vast majority of the cranes will then move down to Spain, with a few crossing the Mediterranean on the way to winter in Africa. These huge passage flocks are considered a large proportion of the estimated 270,000 cranes found worldwide.

The origin of the now-resident British population is not entirely certain, but they may well have come from Swedish cranes that were blown off course while migrating south in the autumn; upon arriving in Norfolk they were able to eke out an existence all year round, including during the harsh Norfolk winter, so they remained. In addition to the resident Norfolk flock, at least 100 to 200

Looking like a plank that has swallowed a pool cue, the common crane is an impressive sight in flight!

Cranes

WHEN
February or March for courting cranes, although they can be seen in any month

WHERE
Horsey/Hickling, Norfolk

migrant cranes arrive in the UK every year with Norfolk being almost the closest point to their migration route down though France.

The spontaneous recolonisation process began in Norfolk in September 1979, when two birds arrived quite naturally in the Horsey/Hickling area of the Broads. The habitat in this area is a landscape of rivers, fens, grazing marshes and waterlogged woodlands, which were originally formed in the Middle Ages when extensive areas were dug for peat. Under strict protection, the pair made this area their home for several years before finally breeding at a closely guarded location in 1982. A number of migrant cranes (some of which only stayed temporarily) then joined the resident flock to augment the numbers. This embryonic flock is currently thought to consist of just over thirty birds, with between one and four pairs nesting in most years since 1996. A pair also recently nested successfully on an RSPB reserve in the Suffolk Fens, and there are plans to begin a reintroduction project in Norfolk with surplus eggs sourced from Brandenburg in Germany. The eggs and chicks from this project will, it is hoped, be reared under carefully controlled conditions, before being placed in a release area until they are self-sufficient, when, with luck, they should be able to boost the current numbers of wild birds.

The 120 centimetre tall Eurasian crane is an incredibly distinctive bird and unlikely to be confused with the much shorter grey heron or the differently coloured white or black storks, which occasionally cross to Britain during the spring or autumn. Adult birds have a blue-grey plumage, a crimson-red patch on the crown and a long white stripe beginning behind the eye that stand out against the black head and neck. The wings and tail feathers serve to add substance to its otherwise lean and gangly frame. The crane's long straight neck and outstretched legs, which extend well beyond the tail, give it a very different flight profile to that of herons and geese, and they will often fly in a shallow V formation on their 210 centimetre wingspan, a slipstreaming technique designed to conserve energy while merrily bugling away to each other.

The juveniles initially have grey head feathers tipped with cinnamon, and will take at least three years to develop the full adult plumage through a series of moults. They will only attempt to breed for the first time in their fourth or fifth year. As befits a large bird that takes time to mature and that has a slow reproductive rate, cranes are relatively long-lived, the oldest wild bird being a ringed Finnish crane which survived to at least seventeen years of age in the wild.

Once seen, the elaborate courtship of the crane is never forgotten, and, although their delightful behaviour can be seen at any time, it is at its most intense prior to the onset of the breeding season in February or March. Cranes will generally mate for life and an established pair will reaffirm their bonds with much 'unison calling', which involves a complex series of coordinated calls that has been heard from as far away as three and a half miles. During this procedure the pair adopt a very specific posture and then proceed to throw their heads backwards, with their bills pointing skywards while trumpeting to one another. The male is usually responsible for initiating this display; he calls much more

than the female and additionally often lifts his wings over his back in heraldic fashion, while the slightly more conservative female keeps her wings down by her side. All the cranes become regularly engaged in a good deal of dancing, bowing, jumping, stick tossing and wing flapping, while delivering their wonderful bugling call. This highly ritualised behaviour also occurs with juveniles and between unpaired birds in the flock and is thought to be used to assert rights, relieve tensions and strengthen family ties.

The crane's nest consists of a heaped mound of wetland vegetation, usually in or around pools, on to which two eggs are laid by the female in April or May. Both birds share the incubation, with the eggs hatching at around 28 to 31 days and the chicks generally hatching 48 hours apart. The male takes the primary role in defending the nest against any potential predators, with any young generally fledging at around 65 to 70 days. The fledging rate of chicks in the Norfolk flock has been very low compared to other populations in northern Europe, and it is thought that egg predation by anything from foxes to mink could be one of the main factors holding back the population.

Food is not thought to be a problem on the Broads as the omnivorous cranes feed on shoots of grass, seeds such as acorns, the leaves of a whole variety of crops, potatoes, insects of all kinds, frogs, snails, worms and occasionally even small mammals and birds. Plants are thought to be particularly important during the winter months when other food items become scarce.

While not currently threatened on the Continent, certain crane populations may well be under threat if the loss and degradation of breeding habitat continues, or birds are still shot or poisoned if they damage freshly planted crops. As the Broads are one of our newest national parks and the largest protected wetland, there is hope that the cranes will have plenty of habitat to choose from, if and when the population really takes off. With large areas of fenland only a few wing flaps away and more areas being created as part of the Great Fen Project, there are grounds for optimism that there will be the space, peace and solitude that this wonderful, yet shy and retiring, bird truly deserves.

22 Hunting peregrines

The beauty, power and sheer presence of birds of prey have captured the imagination of man for thousands of years, and none more so than Britain's largest resident falcon and undoubtedly our most impressive hunter, the peregrine falcon. The peregrine has been used in falconry since the time of the nomads in central Asia in 1,000 BC, but is also the world's most cosmopolitan bird of prey, with 17 recorded sub-species breeding on all continents apart from the Antarctic and in most habitats, with the exception of the ice cap, extreme deserts and tropical forests.

The peregrine's Latin name of *Falco peregrinus* translates as 'wandering falcon' and refers to some of the more northerly populations, which breed up on the Arctic tundra and migrate south after the breeding season to avoid the harsh winters. In areas with a much milder year-round climate like Britain, the peregrines are permanent residents, either remaining at the breeding site all year, or moving just a small distance to the coast in the autumn.

As with many birds of prey, the female or 'falcon' may be as much as 30 per cent larger than the male or 'tiercel', an English corruption of a 16th century French term meaning 'one third' – which accurately translates the size differential of the sexes. The heavier female weighs between 900 grams and 1,300 grams and is about the same size as a carrion crow. Both sexes have very similar plumage and, when seen perching, they are very distinctive and could be confused with nothing else apart from the more diminutive falcon, the hobby. The adults have blue-grey upper parts with barred white under sides and a black head and 'moustache', which contrast with their white neck and throat, giving the birds a hooded look. The cere – a patch of waxlike skin – around the eye and huge feet are bright yellow, with the juveniles mostly looking browner and more heavily streaked.

Traditionally, most British peregrines breed in mountainous craggy outposts or coastal areas with a westerly and northerly bias, but an increasing number have recently begun nesting in urban areas, where man-made structures such as power stations, pylons, high-rise flats, cathedrals and churches are, in many ways, a perfect replica of the original cliff ledges where most pairs still breed. Apart from man, and possibly eagles in northern and western Scotland, peregrines have no natural predators so, providing they make it through the first tough year of life when they have much to learn and a mortality rate reaching 70 per cent, their chances of surviving up to 15 years in the wild improve considerably.

Although the peregrine will very occasionally take small mammals, given the choice, it feeds almost exclusively on a whole range of small to medium-sized birds. Peregrines are essentially a supremely opportunistic species, taking the

LEFT: Just how fast a peregrine stoop actually is, is a subject of much conjecture; however, there is no doubting its sheer aerobatic grace and manoeu-vrability as it closes in for the kill.

RIGHT: With plenty of cliff-like breeding opportunities and plenty of food, it is not surprising that many peregrines have moved lock, stock and barrel into our cities, none more so than this pair in Baker Street in London.

bird prey that may well be most abundant locally; they have been recorded taking an astonishing 98 species in Britain! In upland areas, birds such as meadow pipits and red grouse figure prominently in their diet, while, for coastal cliff-nesting peregrines, seabirds are the obvious choice during the breeding season before the falcons move on to the estuaries in winter to hone in on waders and wildfowl.

In cities, although feral pigeons and starlings are unsurprisingly taken in abundance, other birds regularly identified from their grisly prey remains include blackbirds, greenfinches, snipe and teal. Although most peregrines hunt primarily at dawn and dusk when their prey is at its most active, many of the urban peregrines may well also be carrying out a substantial amount of nocturnal hunting with help from the city's lights. This theory is supported by the wings, legs and feathers of rare migratory species, such as roseate tern, corncrake and black-necked grebe, that have been found at a number of urban peregrine sites across southwest England and must have belonged to migrating birds moving north to their breeding grounds under the cover of darkness. The size difference between the peregrine sexes also means that a breeding pair will frequently be able to exploit a wider range of prey. This is particularly important when both parents are under pressure to keep hungry chicks fed, as the falcon can focus on larger, heavier birds like wood pigeons, leaving the tiercel to concentrate on smaller birds such as snipe and blackbirds.

Although not the quickest bird in level flight, the peregrine is considered to be the fastest animal on the planet during its legendary hunting dive, otherwise known as 'the stoop'. This involves the falcon plummeting downwards at a fairly steep angle, with its wings and tail folded in and its talons tucked away, to strike its victim from above at an incredible speed. The prey is first spotted from either a well-used perch or while in flight, and the bird either takes off, or, as

OK, so these aren't peregrines, but, if you fancy another bird of prey spectacular, then you could do much worse than visit the red kite feeding station at Gigrin Farm in the heart of Wales.

Peregrines

WHEN
Although they can be seen hunting in any month, May to July is the easiest time to watch them hunting

WHERE
Numerous locations including Avon Gorge, Bristol; South Stack Reserve (RSPB), Anglesey, Gwynedd; Derby Cathedral, Derbyshire

the peregrine needs a considerable height advantage in order to enact the stoop, rises even higher as it climbs on pumping wings – sometimes even in the opposite direction to its prey – until it reaches the necessary height and position. The bird then accelerates towards its quarry and, if it misses, will throw up its wings in such a manner as to redirect its flight immediately upwards at the same trajectory as the angle of descent, using the momentum gained to rise above the prey again in preparation for another swoop.

After the peregrine has struck the bird with its talons, it will either swing around and pluck the bird out of the air, or fly down to retrieve its victim, which will have fluttered or crashed down to the ground. Prey that is not killed immediately by the initial strike is usually dispatched by a special notch in the peregrine's bill, which is designed to dislocate the bird's neck. The item is then usually carried back to a favourite plucking post to be denuded of its feathers before consumption. Most of the carcass will be eaten, apart from the head and wings, with any feathers and bones that cannot be digested later being regurgitated as a pellet.

The actual speed achieved during a peregrine's stoop is one of those questions that has never been satisfactorily answered, with estimates ranging from a very extravagant 250 mph to a much more conservative velocity of 89 mph as clocked by radar. Their speed will, of course, depend on a whole variety of factors such as air temperature, height above sea level and air pressure, but either way it seems that they regularly and comfortably break the UK motorway speeding limit! In order to avoid damaging their lungs with the pressure from diving at such speeds, they have small bony prominences, or tubercles, in the nostrils to reduce both the shock waves of the air entering their passages at speed and the change in air pressure, enabling the bird to breathe more easily while diving. The nictitating membrane – an inner eyelid – is also drawn over the eye to protect it from the wind, while still allowing clear visibility to make an accurate strike.

Once caught, the hapless victim is mercilessly plucked before being taken to the nest to be fed piecemeal to the chicks ... what a way to go!

Peregrines reach sexual maturity after one year but they will rarely breed until after two or three years. Successful pairs often mate for life and show great fidelity to their territories by returning year after year, even though the precise nest location may rotate or vary each season. Before the onset of breeding, the pair will often renew their relationship, initially with cooperative hunting, which involves catching a number of prey items as a team. This behaviour will then quickly move on to dazzling courtship flights, where the pair will chase and stoop each other, before pulling up, rising and repeating the same actions again in a manner that is only a slight modification of their basic hunting techniques. Bonds between the sexes are then finally cemented by food-passes between the male and female; the falcon will frequently rise up to meet the food-laden tiercel in mid-air before inverting underneath him to grasp the prey from his talons. The female is then left to choose the nest site before mating and laying begin.

As the courtship develops, the pair become increasingly aggressive in defence of their nest site, particularly against other peregrines; an intruder will invariably be driven away by the incumbent pair with their warning calls ringing in its ears. The nest itself is often nothing more than a shallow indentation or scrape on a sheltered cliff ledge, or, in the case of city peregrines, in a convenient nook or cranny tucked away on their artificial cliff of choice. By March or April the female will have laid three to four whitish eggs with reddish-brown blotches, which she will incubate for 31 to 33 days with occasional help from the male during the day. On hatching, the huge-footed chicks are covered in white down. Initially both they and the brooding female are reliant on the male to bring food to the nest from the pair's hunting territory, which can extend from close to the nest to as far away as 26 miles. On delivery of a carcass from the subordinate male, the female will then take it over and proceed delicately to feed morsels to her chicks. As the chicks increase in size on such a protein-rich diet and are able to feed themselves, this leaves both parents free to hunt, and the surviving chicks will normally fledge at around 39 to 40 days from hatching. Despite having taken to the wing, the fledglings remain dependent on their parents for at least two months afterwards while they slowly learn the fine art of hunting.

Falcons became endangered from the use of pesticides, particularly DDT, between the 1950s and 1970s, as biological accumulation of the organochlorines in the birds' fat tissues led to either a thinning of their eggshells or higher incidences of infertility of the eggs themselves. Since the banning of these chemicals and the increased protection of nesting locations, the population has recovered and the falcons have reoccupied many of their traditional breeding haunts. Additionally the move into cities worldwide has been a bonus, with an astonishing 18 pairs of peregrines now breeding on Manhattan Island in New York. With the current British population thought to be around 1,400 pairs, with at least 65 pairs currently breeding on man-made structures in urban environments, this time, it is hoped, this charismatic bird is here to stay.

Spiders' webs

WHEN
June to November, with webs
showing clearly on early
autumnal mornings

WHERE
All habitats, with gardens
being as good as any

Spiders' webs

While many wildlife spectacles may require a journey, timing and, of course, the requisite degree of luck, surely one of the most amazing animal feats to be seen in Britain can be commonly encountered no further away than in our own back garden. Love them or loathe them, spiders have colonised every British habitat, and, while many spiders spend their lives out of sight, the fact that our countryside is studded with their architecturally stunning cobwebs is a constant reminder that we live in a world dominated by our eight-legged friends.

With over 40,000 species known today, spiders are a phenomenally successful and diverse group, ranging from the bird-eating spiders of the South American tropical forests with a leg span of 250 millimetres, to the smallest-known spider from Samoa, which, at 0.5 millimetres, would fit comfortably on the head of a pin. Even in Britain, which has comparatively few spiders and a well-studied fauna, there have been over 50 new species discovered in the last 50 years.

Irrespective of their huge diversity, all spiders have one strand in common: silk. While all spiders will produce silk in some form, the group that has taken silk weaving to the finest art and leaves the most recognisable and distinctive webs in Britain are the orb-web spiders, a large family that spin, more-or-less-circular webs inside a framework. From the moment they hatch, the orb-web spiders, such as our common garden spider *Araneus diadematus*, begin a life of almost continuous silk production.

Most spiders are able to produce between three and six kinds of silk from different silk glands in their abdomen, with the orb-web spiders having the widest array of threads at their disposal. Each set of glands – which are connected by ducts to one or more spinnerets – are responsible for producing threads with very different functions. Different types of silk threads include those needed for the frame of the web, prey wrapping, or even the special silk glue for the sticky threads. Most spiders have three sets of small and inconspicuous spinnerets, which, in turn, have a number of fine tubes or spigots that act as the openings of the silk glands. These spigots can be of different sizes, and possess valves with which they are able to alter the thickness of the silk being extruded. By increasing the blood pressure around the glands, the syrupy liquid-silk starts to exit the spigots and, as the silk is not able to be actually pumped out, it has to be pulled out by the legs, a process that can be seen very clearly when a spider is wrapping up a prey item. Originally it was thought that the air caused the silk to harden, but it is actually the act of pulling the silk by the spider; this leads to the protein molecules in the silk being stretched, which causes them to bond together and solidify as strands.

So adept are the orb-web spiders at managing their silk that they are capable

Norman Foster would have a job on his hands designing anything approaching a cobweb's strength and beauty!

of spinning at least two different types of silk at the same time from different spigots and are also able to successfully manipulate these separate stands with their legs and spinnerets so they can be simultaneously used for different functions. The individual silk strands in orb-web spiders can be a little as 0.000015 millimetres in diameter, although several strands may well be fastened together to form

Despite being incredibly thin, spider silk is considered to be the strongest of all natural fibres; it is twice the strength of a steel strand of an equivalent thickness.

a functional thread, and the average thread size by the garden spider will be no more than 0.003 millimetres wide. Infinitesimally thin fibres are measured in terms of 'denier', which is the mass in terms of grams of the thread per 9,000 metres: compare the average human hair that measures around 50 denier with the silk of the garden spider at 0.07 denier.

Silk is the strongest natural fibre but it is also incredibly elastic, as can be easily demonstrated when watching a finished web billowing in the wind without damage. Most spider threads are thought to be able to increase their length by up to a third without breaking; a steel thread, by comparison, will snap at eight per cent more than its original length.

While many web-spinning spiders will simply add silk to their webs from time to time in order to increase its size or carry out a repair, the orb-web spinners such as the British garden spider will break down and re-form a new web every night. An average garden spider's web may contain as much 25 metres of silk, which, although incredibly light, still represents a considerable investment of its time, energy and raw materials, so the spider will firstly ingest all the old cobweb, then liquefy and recycle it through the glands in the form of fresh silk. Despite the web being 40 to 50 centimetres across, the garden spider will then construct an entirely new web, containing thousands of connections, often in less than an hour. Once finished, the web will be able to trap and ensnare prey weighing several thousand times as much as the web itself.

Apart from man and a few very specialist insects, spiders are the only animals to set traps to catch their prey, and the orb-web cobwebs have been perfectly honed over millions of years to catch the maximum amount of prey with the minimum effort. In all, the construction of an orb-web has to be one of the most amazing technical accomplishments in the animal world.

The first, and often most difficult, step for an orb-web spider when building a web is to establish the bridge thread from which the finished product will be suspended. The spider either produces a dragline and then uses the wind to carry this first adhesive thread over to a suitable surface such as a leaf or a twig where it will adhere, or, alternatively, attaches the silk to an anchoring point and then pulls a thread out behind her as she explores the surroundings for another anchorage point. The spider will then test the safety of this tightrope as she walks along its length while strengthening it with more layers of silk.

With this bridge secure she will then form a second thread which will hang down loosely from the first taut thread like a washing line, and from which she will proceed to lower herself from the middle of the slack thread until she reaches a point where this vertical strand can be secured and pulled to form a Y shape. The centre of the Y then marks the centre of the web and all the three arms represent the radials or spokes of the web. By scuttling up and down these three radials, constantly producing silk, the spider will now be able to form the rest of the framework and add the rest of the radials, which are very precisely spaced as the spider uses its body to measure the distances. Adult spiders produce fewer radials than young ones, but generally the number of these lines running from the centre of the web tends to be species specific.

The spider will then move to the centre of the web and ensure the radials stay in place by producing a spiral of non-sticky, evenly spaced silk as she works her way to the outside of the web; this will also be used to enable her to move around the web during its construction. Then, starting from the outside, she will create the adhesive spiral threads, or 'catching spiral' by combining drag-line silk with a viscid silk produced by another gland, which, when stretched, breaks up into little tiny droplets strung out along the thread. These droplets of glue are responsible for transforming cobwebs into bejewelled works of art when they are touched by the morning dew in the autumn.

Using her legs as rulers and the radiating lines and non-sticky spiral as guidelines, the spider is able to fix the silk in place with speed and precision as she bobs around the web like a little pendulum. The garden spider will then finish off by weaving a series of short irregular strands of silk in the centre to create a disc where she can wait for dinner. A spider positioned in the centre of the web can sometimes, however, make for highly visible prey for birds so many orb-web spinners will try to reduce the risk of predation by hiding at the edge of the web with at least one foot still in contact with the web to sense any vibrations.

As soon as an insect blunders into the web, the spider rushes towards the vibrations, making sure to walk only on the non-sticky radials where possible. If the prey is small, such as a fly, it is often bitten and eaten there and then. Larger prey items up to the size of the spider itself are bitten and then quickly wrapped with special silk to ensure that as little damage as possible occurs to the web. It has been calculated that the web itself will need to hold a victim for at least five seconds in order to enable the spider to reach the prey, inflict a bite and then wrap it. The near-vertical orientation of the orb-web cobwebs means that prey will rarely fall straight through, but are likely instead to be ensnared by the next sticky strand of the spiral below.

While arachnophobia must rate as one of the most common, and often irrational, fears, it is worth bearing in mind that no spiders in Britain are capable of inflicting a poisonous bite; spiders also kill a vast number of aphids and other pests, making them the gardener's friend. So, the next time you see cobwebs adorning your rose bush, remember that the innocuous spider is a much better insecticide than any spray money can buy.

Puffins

WHEN
End of April to beginning of
August

WHERE
Skomer, off West Wales;
Farne Islands, off Seahouses,
Northumberland; Orkney and
Shetland Isles

20 Wheeling puffins

There must be very few British people who cannot identify the iconic image of a puffin. However, despite these being one of our most common seabirds, only a tiny percentage of the population will have seen these characterful little seabirds in the flesh because of the remote nature of their breeding haunts, and it is a red-letter day for birdwatchers when they see their first puffin, with its handsome looks and engaging personality.

But the puffin is also a very sociable creature and it is premature to say you have witnessed the real 'puffin experience' until you have seen them interacting in profusion near their burrows or wheeling in huge flocks around jagged sea cliffs.

The puffin is undoubtedly the best-known member of the auk family and certainly the most compact and dapper, with its diminutive upright stature, pied plumage and seemingly overgenerous use of eyeliner. The puffin is far from being just monochromatic though, as its black and white plumage contrasts markedly with its vivid orange legs and feet and the most extraordinary coloured bill, which is compressed sideways and becomes daubed with smoky-blue, buttercup yellow and orange panels in summer. This bill is the ultimate multi-tool, as it is used in courtship, for fighting, for digging and for catching food. It is also used by puffin biologists as virtually the only way to separate the largely identical sexes – the bills are generally noticeably bigger on the males.

In common with all the other British breeding auks, puffins nest colonially, but, unlike guillemots and razorbills, for example, which use their legs on dry land as little more than landing gear or for standing on, the puffin enjoys a good stroll. It is not uncommon to see puffins dashing about the burrows like little clockwork toys, whether it be to greet an incoming mate, or to go and investigate the commotion when a row breaks out among neighbours. Despite being very competent walkers, however, puffins are not the most distinguished of flyers, as their short, stubby wings have evolved to be a compromise between enabling them to fly to and from their feeding areas and allowing them a means of propulsion under water when chasing their prey. By flapping their wings at an astonishing 300–400 beats a minute, puffins can fly surprisingly fast and have been clocked at speed of 50 miles per hour. However, they are not the most manoeuvrable of flyers as, when they are forced to land downwind, their technique of high-speed stalling can occasionally lead to inelegant crashes on land or sea.

The reason why so few people have actually seen a puffin is down to the fact that the birds mostly live on isolated islands or remote and inaccessible locations on the mainland. This is because two essential prerequisites for

The collective noun of a 'circus of puffins' does not do justice to surely our most enchanting, sociable and most easily identified seabird.

successful puffin breeding colonies are that they must be free from disturbance and from land-based predators such as rats. Where these conditions are best met and the sites are also close to good feeding grounds, populations can number in the tens of thousands.

The puffins only spend a mere five months each year at the colony, with the rest of the time spent on the high seas riding out the autumn and winter storms in small groups. Early spring sees them freshly moulted ready for the breeding season and also marks the time when they begin to return to their colonies. Upon arrival, they initially spend most of their time on the sea close to and below the burrows and seem very reticent to come ashore. But, as numbers build up, steadily larger and larger flocks take to the air and wheel spectacularly over their respective breeding sites in a distinct circuit, which enables them to check out the lie of the land safely in numbers before returning to the security blanket of the water. As the urge to breed becomes stronger, small groups will then touch down on land before gingerly walking over to inspect their burrows. Once the first brave souls land this encourages the rest to join them and numbers around the burrows quickly swell.

Puffins generally remain faithful both to their mate and burrow location from year to year; as it is doubtful that pairs spend the winter together, the emotional reunion probably occurs at the nest site. As more and more puffins come ashore and more pairs become reunited, courtship becomes the order of the day. Each pair cements their bond for the oncoming breeding season with a spot of 'billing', which involves the gentle broadside slapping of one another's bill. This behaviour can often become a spectator sport with nosy neighbours coming over to watch the spectacle.

The vast majority of puffins nest in earthy burrows and established pairs remember the exact location of their underground den and are prepared to defend it against other puffins with no fixed abode. Possession seems to be nine-tenths of the law in puffin society, and trespassing puffins will be warded off by aggressive open-beaked and winged postures by the owner. If this fails, a fight can ensue where the puffins lock bills in an attempt to assert dominance; early in the breeding season this can easily escalate with locked-together puffins tumbling down the slope into the sea at sites where there is a real shortage of burrows. In addition to protecting their nest sites, a good deal of housekeeping takes place upon arrival with the paired-up puffins making frequent furrows inside to clean them out and repair any damage that has occurred over the winter. The bill and feet make excellent digging and shovelling tools for individuals keen to carry out some impromptu maintenance, or, in some cases, where pairs need to start a burrow from scratch. The end product will be anywhere from one to two metres in length and usually parallel to the surface.

When close to mating, there is much exaggerated head tossing by the male

and rapid vibration of his wings as he impresses his partner and asserts his position in the colony; the brief carnal act itself usually occurs at sea. The single large egg is laid in a grass and feather-lined nest chamber at the back of the tunnel during the first or second week of April, which is very early compared to other British seabirds. Both sexes incubate the egg in turn for a period around 40 days, and undoubtedly one of the highlights of the summer is seeing the first parents carrying fish ashore indicating that their chick is at the vanguard of a flurry of hatchings.

Puffins socialising of
an evening before
bedding down for the
night on the sea.

The puffin young – delightfully called 'a puffling' – is a dark, fluffy ball of downy feathers. The puffling will stay out of sight, safely in the burrow for the entire duration of its adolescence. For the adults, chick rearing is a frenetic period as both sexes make prolonged trips out to sea in order to satisfy the needs of their growing hatchling. The most common fish brought back by the adults tends to be sand eels, but sprats and small herrings feature in the puffling's diet too. Of course the snap that every puffin photographer wants to get is that of an adult with a beak crammed full of fish, making it look like a film extra with a huge Mexican moustache straight out of the spaghetti westerns! The puffins are able to hold, chase and catch other fish without dropping fish already caught as the held fish are trapped between a series of spines on the roof of the mouth meaning that puffins carry back, on average, 4 to 20 items in their beaks, with the record being an incredible 62! A popular misconception about the way puffins carry the fish was that they were neatly arranged in the beak, with heads and tails alternating on each side, but this seems to be untrue and the fish are held in a haphazard fashion.

While the adult puffins spend most of the day actively feeding their single young, the evening is a time when the puffins gather on the slopes outside their burrows, so this is also a good time to catch the best behaviour as pairs and neighbours interact in a seemingly very sociable manner prior to leaving the pufflings safe in the burrows and settling down for the night on the sea.

Most fish are caught within a few kilometres of the colony and then the puffins fly back as quickly as possible to the colony to avoid having their food stolen by piratical bird species keen for an easy meal. While only the greater black-backed gull will regularly kill and eat an adult puffin outright, there is a veritable queue of thieves happy to steal any fish dangling in an enticing manner from a returning puffin's bill. Arctic and great skuas (at the northern colonies) and herring and lesser black-backed gulls (at most breeding spots) are all birds that will attempt to chase and harass the puffin into offloading its hard-won fish.

Despite the adults having to run this daily gauntlet, several feeds a day will usually ensure the puffling will be able to fledge between 38 and 44 days after having been hatched. Just prior to fledging the feeding tails off quite dramatically and the chicks will even lose weight before finally sticking their head out of the burrow for the first time. The puffin fledglings leave the burrow alone and under the cover of darkness to minimise the risk of predation from large gulls eager for the easy meal of a naive young puffin, and they get no further help from their parents once they depart for the sea.

By the end of August the last stragglers have headed for the sea and the nesting sites become puffin-free zones once again. The deserted burrows seem to have something of the feel of a 'shabby coastal resort that has seen better days' about them, until they are once again brought to life by these marvellous little seabirds the following spring.

Autumn

WHEN
September to October
depending on the season

WHERE
Westonbirt Arboretum,
Wiltshire; Bedgbury Pinetum,
Kent; Thorp Perrow Gardens,
Yorkshire

19 Autumnal trees and toadstools

Of all the spectacles, surely one of the easiest to catch is the sublime autumnal colours of our deciduous trees, as well as the fungi on the woodland floor. This fanfare of colour marks nature's final flourish before the long descent into winter and provides a wonderful fillip to any woodland walk. Though the changes in leaf colour are enjoyed by many, the real reason why so many leaves undergo such a radical transformation before falling from the trees is still shrouded in mystery.

While the woody parts of trees can invariably survive the cold, as can the needles of most evergreen trees which are especially resistant to cold and moisture loss, the leaves and stalks of deciduous trees fade, die and fall off each autumn as they become an indulgence the tree cannot keep. Without their leaves, the deciduous trees are forced to enter a period of dormancy, and survive by living off starch products stored up during the spring and summer.

Being perennial beasts by nature, most trees will live for a number of years, and, for species living in temperate regions such as Britain, this means they will have to undergo the annual period of sufferance that is winter. During this time, light levels often become too low to photosynthesise and the tree also has to conserve water, which is freely passed off as vapour during the summer, but becomes more difficult to obtain when the ground is frozen.

Photosynthesis literally means 'putting together with light' and is the process by which plants convert inorganic substances such as carbon dioxide and water through a complex reaction into organic substances such as sugars. This reaction occurs thanks to the presence of chlorophyll, quite possibly the world's most important compound, the green-coloured pigment that gives most plants their colour and the countryside its hue. Chlorophyll is green as it absorbs red and blue light, but reflects the green part of the spectrum, hence that is the colour our eyes pick up. More importantly, it is the compound that traps that 'most elusive of all powers' – the sunlight – and converts its energy. Chlorophyll, unfortunately, does have the disadvantage that it is not a particularly stable compound, and, ironically, the bright sunlight, which it is able so effectively to harness and use, also causes it to break down. So, in order to maintain an abundant supply of chlorophyll in their leaves and ensure the continued production of sugars, the trees are obliged to synthesise its production continually during the spring and summer months.

The environmental triggers in autumn, which instigate a series of chemical changes in the trees resulting in the leaves changing colour before dropping off, are the shortening days and cooler nights. Well before the leaves fall, the tree

The golden hues of the beech as it turns towards winter are as fine as any of the gaudy reds offered by the North American and Asian maples.

prepares to cut each leaf loose by forming an abscission zone across the leaf stalk, which consists of a special layer of cells that gradually form a seal, before eventually severing the tissues supporting the leaf. When the leaf's weight becomes too much to bear, or it is caught by the wind, the leaf is then shed.

Before the leaf is actually severed, important substances such as nitrogen – which is one of the vital ingredients used in the composition of chlorophyll – are transported out of the leaf so they can be reused next year. As the water and nutrient supply to the leaf steadily becomes cut off along the abscission zone, the leaves cease the production of chlorophyll and the remaining pigment begins to break down with the colder temperatures. Other more stable pigments, which have been present in the leaf but have been masked by the chlorophyll, begin at this stage to show their true colours. The two main pigments initially responsible for the colour of the leaves once the green disappears are the yellow xanthophylls and orange and red carotenes, which are the same pigments responsible for the colours of bananas, egg yolks and carrots. During the spring and summer, both these pigments are responsible for the absorption of sunlight of slightly different wavelengths to chlorophyll before passing it on to the green pigment to use in the production of the sugars.

The colours of some species' leaves are enhanced by further chemical reactions which produce a pigment called anthocyanin which is naturally scarlet or purple by nature. The anthocyanins are manufactured when sugars that have remained in the leaf, or are produced by the last remaining chlorophyll, reach a high-enough concentration to trigger a reaction with proteins in the cell sap. Anthocyanin is responsible for the red skin of ripe apples and the purple skin of ripe grapes. The production of anthocyanin in the leaves of trees with a propensity for going red, such as the North American or Japanese maples, is strongest following a run of warm, dry and sunny days followed by cool nights devoid of frosts, as the sugars become more concentrated in the leaves.

So, if anthocyanin is mixed in with the xanthophylls and carotenes and the weather conditions are suitable for their production, it is likely that the full palate of shades from green, to yellow, to orange and red will be seen when the pigments are mixed in different amounts. In Britain, many of our native species such as hazel, birch and oak tend to be dominated by the carotene and

ABOVE: The magpie ink-cap is a handsome species that, bizarrely, seems to smell of bitumen.

xanthophyll pigments – hence a predominantly yellowish colouring once the chlorophyll has faded in autumn – but beech, sycamore and cherry have the propensity to produce anthocyanins and will go more reddish if the right environmental conditions prevail before the leaves fall.

There are a number of theories as to why anthocyanin, which is energetically so demanding for the tree to produce, is manufactured at all. It seems that the pigment may enable the trees to keep their leaves for a short while longer by protecting them against frost damage, meaning the leaves are able to shunt more of the important compounds out of the leaves before the abscission zone finally cuts off the leaf. Another hypothesis is that insects (like aphids) are not attracted to plants with red leaves and so will lay their eggs – which will, of course, hatch into leaf-munching larvae the following spring – on other leaves and trees thereby saving those that are redder.

Either before or after the leaves fall, they will begin to turn brown as these more stable pigments eventually break down. The tree also has the ability to use leaves as a suitable repository for the dumping of any toxic waste products, knowing that they will soon be cut adrift. The fallen leaves will then ultimately become broken down themselves by soil microbes, invertebrates and fungi, leading to the sequestered minerals and nutrients being recycled, possibly even by the same tree, to put on a show the following autumn.

Should you tire of craning your neck up to check out the colours, autumn is also the best time to catch the parade of toadstools. The subterranean fungal mycelia (or fungal threads) will have had a long, warm period in which to spread out through the soil in the summer and, with the arrival of wetter weather in the autumn, the damp humid conditions created are ideal for the fungi to fruit.

Fungi are actually the cornerstone of many woodland ecosystems as the majority of the large mushrooms and toadstools in woodlands are intimately linked to the trees that surround them by special mycorrhizal associations, which is mutually beneficial to both the trees and the fungi. The unseen filaments of these mycorrhizal fungi will often grow over the surface of the trees' roots forming a sheath, and ultimately result in some filaments actually penetrating cells in the roots. This intimate relationship allows the fungus to obtain most of its sugars from the tree while the tree will be supplied with a range of essential minerals, such as nitrogen and phosphorous, that has been obtained by the fungus from decaying matter, such as leaves, in the surrounding soil. All the fungi species in the important genera of *Boletus*, *Lactarius*, *Russula* and *Amanita* are considered to have very specific and beneficial associations with trees. One of these is Britain's most famous toadstool, the fly agaric. This instantly recognisable red-and-white toadstool is the subject of countless children's stories, and has a mythical status because of its hallucinatory powers; it is also a fungus that is never found far from its mycorrhizal associate, the birch tree. So, if trees or toadstools are your passion, if you go down to the woods in the autumn you will be sure of a big surprise.

18 Wild orchid displays

It's official – orchids are very sexy plants; no other British plant family can even approach their beauty, glamour or charismatic appeal. In addition, they are the most highly evolved group of plants, and the largest plant family, with an astonishing 25,000 described species worldwide. In terms of diversity, the family reaches its zenith in the tropics, with many species living an epiphytic – or parasitic – lifestyle high up in the canopies of rainforest trees. Although there are only 56 species of orchids in Britain, all of which are rooted firmly to the ground, they are nevertheless a very proud part of our natural history heritage and, when seen in profusion, can justly be described as the perfect summer spectacle.

Despite the comparatively meagre number of orchid species native to Britain they are still a surprisingly varied group, ranging from the diminutive and finger-sized bog orchid, to the stout woodland specimens of the broad-leaved helleborine, which can reach over a metre tall, and from the pale and sickly-looking bird's-nest orchid, to the vibrant gaudy lipstick pink of the marsh orchids. British orchids also show massive disparity in abundance, with the well-named common spotted orchid being found virtually everywhere (with the exception of the Scottish Highlands), while the lady's slipper orchid was reduced to just a single plant at a closely guarded location in Yorkshire before conservationists stepped in.

The actual word orchid comes originally from the Greek *orkhis*, meaning testicle, and refers to the underground tubers, which can clearly be seen on dug-up specimens of species such as the early purple orchid. The ephemeral beauty and supposed rarity of orchids have led to them being much sought after by botanists and gardeners; as a result many populations were drastically reduced in the 'pin it, pickle it, stuff it' Victorian era. During this wanton period, collectors were certainly responsible for the downturn in numbers of some of our rarest orchids, and possibly the final extinction of one species, the summer lady's-tresses.

The obsession with collecting, and its connection to the disappearance of many orchids, is, however, something of a red herring as the real issue has been the wholesale destruction of huge numbers of sites. This has resulted in populations simply vanishing under tarmac and concrete, or the irrevocable degradation of many populations as a result of either poor farming or forestry practices. Even today, many of Britain's best orchid sites are still under threat either from nutrient enrichment from fertilisers or a complete lack of under-standing of the needs of orchids even at protected sites.

Even plant novices are able to recognise orchids thanks to the popularity of cultivated tropical varieties that adorn our homes, but giving a precise definition

LEFT: With its rather rudely shaped tubers below ground, it is no surprise that the early purple orchid has the Latin name Orchis mascula.

RIGHT: Looking like its uppermost flowers have been scorched, the burnt-tip orchid also has a surprising scent of stewed cherries.

is not straightforward without resorting to botanical terminology. All British orchids can be described as perennial herbs without any woody parts, and with simple leaves that are arranged alternately around the stem. The flowers are carried in a single spike near to the top of the stem with the male and female parts of the flowers fused together. Each flower is composed of three sepals (modified leaves) and three petals, with one of the petals forming a lip, which can often be brightly coloured and intricately patterned, and which extends backwards to form a pouch or spur, often holding nectar.

In addition to their five-star appeal, orchids have fascinating and mercurial lives, though there are many gaps in our understanding of their ecology. It is still poorly understood why certain species flower in profusion one season and then scarcely appear the following year, or why common twayblade orchids, for example, may take as long as 20 years before they flower. Undoubtedly fundamental to the survival and success of orchids is their special relationship with fungi, which they need in order to germinate and grow. Many orchid seedlings will spend years below ground, during which they are wholly reliant on fungi to provide them with the nutrients needed to sustain life until they are able to generate their own food by photosynthesis (for which they need light). Some orchids are thought to be so exacting that they are only able to germinate and grow in the presence of just one species of fungus, which may account for the natural rarity of some orchid species, as they will be strictly confined to locations in which the fungus is present.

Orchids

WHEN
From late April to August;
June is the prime month

WHERE
From deciduous woodland to
marsh, fen, grasslands, heath
and moorland, depending on
the species. Park Gate Down,
Kent (Kent Wildlife Trust);
Kenfig National Nature
Reserve (Countryside Council
for Wales)

Stumbling on a rash of orchids, like these 'common spotteds', will surely mark the highlight of any summer walk.

Despite the lip-mimicking, and the fact that it smells like a female bee in order to dupe male bees to spread their pollen, it is ironic that the vast majority of these bee orchids resort to self-pollination.

As the growing conditions orchids need to germinate are so particular, including the presence of the right fungus in the correct type of habitat, orchids have evolved to produce a huge amount of seeds to maximise their chances of at least one seed landing in the right place. It is believed that the greater butterfly orchid, for example, may contain as many as 25,000 seeds per capsule, and, while the seeds are so small and light that they can easily be dispersed long distances by the wind, the trade-off is that they contain very little in the way of food resources, so linking up with the correct fungus as quickly as possible is crucial for their survival.

The fungi that associate with the orchids are known as mycorrhiza, a specialist group that live in the soil, often in and around the roots of a huge range of plants. As the orchid seeds are unable to photosynthesise, the fungi provide them with nutrients obtained from the decomposition of surrounding organic matter and effectively form an extended root system to the germinating seedling as the fungal mycelium spreads out into the surrounding soil to provide for the embryonic orchid. Even when the orchid finally appears above ground and is able to begin photosynthesising following the formation of its first leaves, the relationship with the fungi does not end. While a few species, such as the bird's nest or coral orchid, are incapable of chlorophyll production and are utterly dependent on the fungi all their lives, many orchids probably continue to obtain nutrients from both photosynthesis and fungus in different proportions. This skill to 'tap' the fungi's ability to provide nutrients may explain why orchids are able to spend so long effectively dormant underground even after having flowered in previous seasons, and why they are able to thrive in marginal habitats on poor, thin soils.

Roadside verges can provide surprisingly rich picking for orchids, as is the case with these 'pyramidals'.

However, with orchids nothing is ever quite as it seems. Unlike normal mycorrhizas, where the photosynthesising plant is able to return the earlier favour by providing the fungi with a source of their carbohydrate (a term called mutualism), there is no evidence of any nutrient transfer from the orchids to the fungi. The association is a parasitic one whereby the orchid 'cheats' in its relationship with the fungi as it gets 'something for nothing'.

Orchids have not, of course, evolved their renowned beauty and intricate flowers for our benefit, but to attract much smaller visitors, in the form of pollinating insects, which they need to fulfil the role of sexual reproduction. The whole reason for the orchids' alluring flowers is to enable cross-pollination in order to ensure a mixing of genetic material and the creation of a larger, more vigorous gene pool. However, ironically, many orchids that exist either in scattered populations or survive on the edge of their range, may well also be able to successfully self-pollinate, if cross-pollination has not occurred after a certain length of time.

In order to maximise the chances of out-sourcing genetic material many orchids have evolved a number of ways of attracting insects in the first place and then encouraging them to take the pollen to the next flower. The bright colours and different scents, for which many orchids are justifiably well-known, operate as beacons to attract the insects, with the lip being presented as a convenient place to land. While some British orchids will supply nectar to attract pollinating insects, such as the butterfly orchids that smell most strongly in the evening to attract pollinating moths, many species, like the fly orchid, will attempt to use deceit in persuading their chosen quarry of digger wasps to visit their flowers, which remarkably both mimic and smell like the female wasps. Male wasps drawn in for the chance to 'copulate' with the flower are then stamped with pollen sacs or pollinia, which are stuck to their heads before the sexually frustrated wasps fly off to try their luck, hopefully with another fly orchid, resulting in the pollinia being removed once the wasps make contact with the stigma.

While deciduous and coniferous woodland, marshes, fens, heaths, moors, roadside verges and dunes all have their fair share of specialised orchids and displays, by far the best places to see a true orchid spectacle are grasslands. The chalk downs of Southern England have perhaps the finest array of species and sheer numbers of orchids, with some of the premium reserves holding 13 or 14 species and thousands of spikes for a brief period in early summer. Reserves like Park Gate Down in Kent put paid to the myth that all orchids are rare, with literally thousands of fragrant, pyramidal and common spotted orchids rubbing shoulders with much rarer species, such as the wonderfully named monkey and lady orchids. Of course, as orchids are important indicators of sites of real conservation interest, wherever orchids are present in such profusion, there will be many other rare and interesting plants present, and in close attendance will be a whole range of bees, butterflies and hoverflies, as well as the all important pollinating insects.

Grasslands, if left to their own devices, will develop into scrub and eventually revert back to woodland, but the reason why sites like Park Gate Down produce so many orchids each season is because the grassland is maintained in a state of arrested development. Careful grazing and mowing regimes at the right times each year are crucial to ensurd that more aggressive growing plants are kept in check while the orchids thrive, which will in turn enable visiting naturalists to continue to marvel at this botanical extravaganza.

Whales

WHEN
Mid-June to the end of
September

WHERE
Off the coasts of Mull, Skye
and west Cornwall

17 Whale watching

British seas and the northeast Atlantic Ocean would not, initially, be at the top of most people's chosen locations for watching whales. Historically, these waters were extensively plundered for cetaceans – whales, porpoises and dolphins – in the 18th and 19th centuries, and the unpredictable weather conditions and resultant swells off the British coast can test one's sea legs and make looking for cetaceans difficult at best. But, with whale numbers steadily improving, and their annual movements and behaviour better understood now than ever, providing you don't go out in a gale, a large and very memorable marine mammal encounter could seriously be on the cards close to home.

If you are lucky enough to have an encounter with a whale in British waters, the chances are it will be a minke.

With the world's seas and oceans interconnected, literally anything could, theoretically, sail past the British Isles. The list of whales recorded around the sovereign waters is impressive and includes anything from sperm whales to humpbacks and killer whales. In reality, though, the whale most commonly encountered in British waters is likely to be the minke (pronounced minky), the smallest member of the 'rorqual', or baleen, family to be found in the northern hemisphere. Reaching only seven to ten metres at maturity, and weighing a similar amount in tonnage, the minke is only a third of the length of the world's greatest ever creature, the blue whale; don't worry, though, when one is seen swimming under the hull of a small tourist boat in the waters around Scotland's Inner Hebrides, it will look anything but diminutive!

The minke supposedly took its name from a Norwegian whaler called 'Miencke' who, in the 18th century, had the habit of harpooning minkes after mistakenly identifying them as the larger, and commercially more valuable, blue whale. His shipmates joked that all small whales they encountered should be called Miencke's whales, and the name eventually caught on. The taxonomy of the species is complex, but there are believed to be three sub-species: the North Atlantic minke, which is the form regularly encountered off Scotland; a North Pacific form; and a distinctive race from the southern hemisphere.

In common with all baleen whales, the minke have long, sleek bodies with a distinctive narrow, pointed snout that looks V-shaped from above. At close quarters, a single ridge running along the top of the minke's rostrum (or beak) from the double blowholes to the tip of the snout is also very pronounced. Their upper parts are dark grey, which can often look black from a distance, and which contrasts with the rarely seen whitish under parts and broad, swirling grey chevrons extending up the flanks. The small curved dorsal fin is set well back on the body, but the best diagnostic feature is a broad white band, looking remarkably like a paint splodge, that spreads across the upper surface of each pectoral flipper and which can easily be seen on a submerged shallow whale close at hand. Freeloading crustaceans can also occasionally be seen attached to the bodies of some minkes, with the parasite's head burrowed into the whale's blubber, while the body hangs loose in the water.

The North Atlantic minkes, which are particularly seen off Scottish and Irish coasts, favour the relatively shallow, but colder, continental shelf waters with their rich feeding grounds during the summer months, and then migrate to the warmer tropical waters during the northern winter. Spending periods of time in different marine environments has meant they have evolved to eat a wide variety of prey species: fish, such as herring, capelin and sprat, make up the bulk of their diet in the north; plankton figures more prominently in the tropics.

To enable the minke to capture large volumes of water, each whale has 50 to 70 pleated throat grooves that run from underneath the lower jaw to just behind the pectoral fins and are able to expand when the whale needs to feed. The most commonly observed way of feeding is lunge feeding, in which the whale lunges towards its prey near the surface at high speed with its mouth fully open and the grooves extended. The mouth is then closed, taking in 'both baby and bathwater'. The water is expelled through the 230 to 260 yellowish baleen plates, measuring 20 centimetres in length and situated on each side of the upper jaw; the food is extracted and swallowed.

Other whales will turn up around our coast, although sadly the only chance people will have of a close encounter is if they wash up dead, as was the case with this sperm whale.

When feeding at depth, however, whales spend comparatively little time at the surface and, following a deep dive that may last over five minutes, they will frequently have to spend several minutes at the surface breathing and blowing a number of times to replenish the oxygen supply before resubmerging. When diving, the tail, or fluke, is very rarely revealed but the body will arch to enable the whale to power itself down below. Unlike many of the larger whales, the blow upon surfacing is rarely visible as minkes begin exhaling while the whale's head is still partially submerged, with the blowholes usually emerging from the water at the same time as the dorsal fin.

As is typical with most baleen whales, minkes are rarely seen in large pods and are more commonly encountered alone, unless they are concentrated in a particularly rich feeding area. Interestingly, they often tend to be separated by sex, age and reproductive state, indicating that a complex social structure may exist and leading to the theory of territoriality among minkes, which would be unique among whales. In the North Atlantic, researchers have found that the larger mature females predominate in the early summer in Scottish coastal regions, with the males migrating further north in the open sea, while the immature minkes arrive slightly later and remain further south.

The breeding cycle in the minke whale is believed to occur annually, with conception taking place in the warmer tropical waters in winter or late spring before the migration north to the feeding grounds for the summer. The gestation period of ten months means that the mothers will give birth to their two-and-a-half-metre-long single calves once they have returned to the lower and warmer latitudes the following winter. The fact that mother and calf are very rarely seen together in Scottish waters would indicate a fairly short lactation period of between four and six months, meaning that the calf may well have been weaned and left its mother's side before arriving in the rich Northern waters. The calves are not considered mature until they have reached at least seven metres in length and may not be ready for breeding until between five and eight years old. But, with minkes believed to live beyond the age of 50, providing they are not killed by their one natural predator the killer whale, they will still have a healthy number of reproductive years ahead of them.

While largely silent in the water, minkes will occasionally breach, enabling good views of something other than their backs and dorsal fins. They can also be curious with both boats and divers, and have had a few close encounters, particularly in the area called The Minches between the Hebrides, and the waters surrounding both Orkney and Shetland. Unfortunately, the minke is still the only whale to be hunted commercially under the guise of scientific research work by Norway and Japan, and, while the international trade in whale components is still banned, 400 to 600 whales, which have passed safely through British waters, are killed by the Norwegians every year.

In the event that you don't see minkes, it is definitely worth keeping an eye open for harbour porpoises, the smallest British cetacean and a species that can often be seen in abundance all year round in Scotland's waters. Often reaching no longer than 1.7 metres, the harbour porpoise has a stout, streamlined body, no beak and a small triangular fin, and is often encountered in small groups. At sea they are usually seen only briefly when they surface to breathe, before they dip below the water again in a perfect arc, as though they are attached to a vertical spinning carousel. Their main food consists of herring, capelin, sprat and silver hake, and their comparatively high breeding rates mean that at least a 100,000 harbour porpoises live around the British and Irish coastline, despite large numbers being killed as 'by-catch' in the North Sea fishing industry.

So, with plenty of tourist boats keen to take out the budding whale watcher, there are no excuses for not getting out on the water and tracking down your very own Moby Dick. And, while you'll certainly not encounter a blue whale with a penchant for a certain Captain's leg, who knows what will be encountered around the beautiful west Scottish coastline?

Grebes

WHEN
In fine weather between
February and April

WHERE
Widely distributed on the
lakes, reservoirs and gravel
pits of lowland Britain

16 Courting great crested grebes

The great crested grebe is undoubtedly one of the most familiar and striking birds to grace our waterways. But this species is not just restricted to being a good-looking dilettante; it possesses an elaborate breeding display that must surely rank among the most beautiful scenes in British nature. The 'great crest', as it is affectionately known to birdwatchers, also holds the distinction of being one of the key species that kick-started the conservation movement in Britain.

While they actually look comparatively drab outside the breeding season, the great crests more than make up for a nondescript winter plumage with an extraordinary summer headdress. Both sexes undergo a second partial moult in midwinter to form a gorgeous frill of chestnut and black feathers or tippets, which surround their white faces like a sundial and are set off by two sets of black plumes on the crown of the head. All these feathers can be erected at will to impressive visual effect in the complex display courtship between the pair.

While male and female look similar, the male is taller, with longer, brighter tippets and a thicker, more elongated bill. Both sexes are equally sleek, streamlined swimming machines and are perfectly adapted for an aquatic life that consists of rearing chicks on a floating nest and fishing underwater.

It is in the depths of winter, after the birds have moulted, that they begin to turn their heads towards the subsequent breeding season. Most great crests choose to pass the coldest months away from their breeding grounds on bodies of water that won't freeze over, such as large lakes, estuaries and coastal waters. Pairing up for the oncoming year can often occur as early as December; having spent a short break apart after the fledging of their chicks from the previous season, the pair will reunite. Displaying begins in earnest around February or March, once the pair have returned to the lake, reservoir or gravel pit for breeding.

One of the main objectives of their famous mutual display is for the pair to secure a territory. Depending on local issues, such as the population density and availability of suitable nesting locations, territories can vary from one extreme to the other: from an area that is only marginally larger than the actual nest site; to a large portion of lake that is easily big enough to find enough food to rear the chicks. The courtship is also of course enacted to cement the relationship and comprises four distinct and highly ritualised elements: 'head shaking'; 'the ghostly penguin display'; 'the retreat'; and 'the weed ceremony'. These different components are often initiated by 'croaking' advertising calls, which are made by birds that have momentarily lost sight of their other half.

Strictly Come Dancing looks clumsy when compared to the finely choreographed movements of the great crested grebes.

Once the birds come together, 'head shaking' is the most common display component and often forms the precursor to the more flamboyant parts of the courtship. The two birds face each other with erect necks and then both alternately and synchronously waggle their heads, or sway them from side to side. This head shaking can then subtly change into a mock-preening movement, where the birds dip their heads down and back so they momentarily give the impression of preening one of the scapular feathers on their back. During the engagement period, head shaking is at its most prevalent and often ends with the birds separating and diving underwater to continue feeding.

Usually after having spent a period apart, the birds will occasionally perform the 'ghostly penguin display'. One bird approaches the other underwater, a fact often betrayed by a line of bubbles and a ripple as it swims towards its mate. While waiting for its partner to surface, the bird sitting on the surface then adapts a special hunched posture, called 'the cat', in which it hunches its head and raises its wings out like a pair of shields and in a manner reminiscent of a displaying cob mute swan. The submerged bird then surfaces close by and, powered by its feet, rises out of the water into a bizarre vertical penguinlike posture with its white belly foremost, in a manner reminiscent of a ghoul.

TOP AND BOTTOM: The pièce de résistance of their courtship is the weed ceremony, which culminates in the pair rising out of the water breast to breast with their weedy offerings.

The ghostly penguin display then often reverts back to head shaking, but, when other pairing grebes may be close by, the pair will often be stimulated to move on to the third component. The 'retreat ceremony' consists of one of the birds dashing off across the water only to settle down on the water and turn to its partner in the cat posture. The stationary bird then seems to be taken by surprise and assumes an aggressive posture with all its head feathers erect, or also adopts the cat posture. As with the head shaking, the ghostly penguin is most commonly performed in the engagement or early courtship period.

The most spectacular of the great crests' rituals is the weed ceremony. This is a feature of the later stages of courtship when the couple have already mated, and is the pièce de résistance of grebe behaviour. After head shaking, the birds will turn and swim away from each other in a formal manner, calling to each other as they go. One bird, followed by the other, will then dive for a period of several seconds before quickly reappearing with a beakful of weed. If they are unsuccessful in locating any weed they will often make do with a twig or even nothing at all, and then they turn to face each other and speed to a central meeting point, with their heads held low and just above the water. As they come together, they rise vertically out of the water, breast to breast, treading water to maintain their position while waggling their heads from side to side. After between five and fifteen seconds they will drop back into the water, discarding their weedy offering, and finishing the spectacular off with a few head shakes.

Both male and female share the roles of three of the ceremonies, with the female initiating the majority of the 'retreats'. Great crests are generally solitary and fairly hostile birds so the origin of these displays is thought to be hostile encounters with other grebes, which have evolved over thousands of years into a ritualised 'contest' in which the male and female are able to recognise one another and thus to establish and maintain firm pair-bonds.

Having consolidated their relationship with a variety of components from the grebes' extensive repertoire, mating then takes place. This usually occurs on a specially constructed platform, which is separate from the nest and has been built from vegetation piled up on submerged plants or tree roots by both birds.

Although not quite as dramatic, an added bonus for the birdwatcher later in the year is when the adults carry their young on their backs. The chicks are precocial – they are quite advanced as soon as they are born – and they often climb on to their parents' backs as soon as they have dried out, to be brooded underneath or between the wings. The chicks look nothing like their parents, and, with their black and white stripy down, resemble animated mint humbugs!

The most recent census indicates a population of over 6,000 pairs of great crested grebes, which suggest the species is now in rude health compared to an estimated population of just 42 pairs in 1860. The main reason for this enormous slump in the nineteenth century was because of the intense demand for the summer head plumes and dense feathers on the grebe's body – otherwise known as 'grebe fur' – to be used in Victorian women's fashion for hat adornments, boas, hand muffs and shoulder capes.

This wanton slaughter was eventually stopped by a series of pioneering Bird Acts, the last of which was passed in 1880; the plight of the great crests also led directly to the formation of the RSPB in 1889. Of course the only places that the great crested grebes should be adorning are our local lakes and gravel pits. A number of the spectacles in this book require skill, persistence and a degree of luck to see, but this really should not be the case with great crests. Grebes are conspicuous birds and are easy to watch in a number of locations where they are used to people. Pick the right location and time of year and, with persistence, you will be rewarded with at least a couple of the components of this famous and dazzling display.

15 Leaping dolphins

There is very little to beat 'messing about in the water', particularly if that water happens to contain some of the individuals from one of only two resident populations of common bottlenose dolphins to be found around Britain's coastline. If you like your animal experiences to be interactive, there is something incredibly thrilling about seeing a dorsal fin cutting through the water as the owner's curiosity proves too much and it comes for a nose around your boat.

If, however, you are a naturalist devoid of sea legs and worried that this might then be an opportunity that could pass you by, fear not, as this is one sea mammal that has such an affiliation with the coast you can often get superb views from a couple of special promontories without even getting your feet wet.

The common bottlenose is the archetypal dolphin in looks. It is the most abundant of all the 34 oceanic dolphin species, and is also the most well-known as it is the dolphin commonly found in oceanariums, and has achieved worldwide exposure as the star in the famous film and television series *Flipper*. The common bottlenose has a wide head and body, long flippers, a hooked dorsal fin and a short, stubby beak or rostrum, which supposedly resembles a bottle and from which the dolphin gets its name. The real nose of course is nowhere near the beak and is situated in the blowhole on top of the head. Bottlenoses tend to be coloured greyer above and paler below, and their one distinguishing feature is that they really have no distinguishing features.

The dolphin's forwards motion through the water is provided by the powerful tail fluke which enables gentle cruising. Like all dolphins, the bottlenose must return to the surface to get an intake of fresh air through its blowhole and will come up to breathe every five to six minutes when feeding at depth, or breathe much more regularly when at the surface. As it is committed to actively undertake regular trips to the surface, it has been suggested that dolphins are very light sleepers and have evolved the ability to turn off half their brain at a time to allow the organ some recuperation.

The bottlenose is also the most cosmopolitan of all our dolphin species and is equally at home in tropical or temperate waters, with a number of coastal populations along the continents and oceanic islands of which two can be found in the British coastal waters of Cardigan Bay in Wales and the Moray Firth off the northeastern Scottish coast. The population in the Moray Firth is the most northerly population of common bottlenoses in the world and they are also the biggest, with the larger, heavier males approaching four metres in length and upwards of six hundred kilograms in weight. This compares very favourably with their cousins in warmer waters, who don't need the bulky body mass to keep out the cold and rarely reach two-and-a-half metres in length.

Seeing a bottlenose dolphin leap clear of the water is enough to make any naturalist jump for joy.

Dolphins

WHEN
The resident populations can be seen in any month, but the weather means they are easier to spot during the summer months

WHERE
Cardigan Bay, West Wales; Moray Firth, Highlands

Both British coastal populations consist of at least 130 dolphins, a fairly accurate figure as most individuals can be identified by experienced researchers from the shape and individual scarring patterns on their dorsal fins. Bottlenoses are intensely sociable and occur in groups. These vary greatly in size, depending on the habitat. Animals at inshore locations form smaller groups of between two and fifteen individuals, with the average in Moray Firth being around eight; offshore pods can number in their tens or even hundreds. These pods are thought to offer better protection from any potential predators and to make it easier to find food, which often tends to be patchily distributed but, when located, is usually abundant enough to feed all the individuals. The composition of the pods also varies, with related females and their juveniles staying together for years, during which time they are visited briefly and occasionally by adult males that have formed strong pair bonds with their own sex.

One of the most thrilling aspects about common bottlenose dolphins is their curiosity around humans. They regularly approach powerboats and often swim effortlessly along on the pressure wave created off the bow as the boat ploughs through the water. It is not just boats that they swim alongside either, as they will regularly surf along on storm-driven waves or even on the bow wave caused by large whales! It is not uncommon to see the dolphins breaching, as they leap clear of the water with a joie de vivre that you would be pushed to see elsewhere in the natural world. The close affinity with humans extends to anyone entering their world, with many divers and bathers being treated to close encounters with particularly playful dolphins. Occasionally dolphins have even rescued injured divers by raising them to the surface, and, in November 2004, a pod of dolphins surrounded and protected three lifeguards from a great white shark for over 40 minutes until they could reach the shore off New Zealand's North Island.

This benevolence does not extend to other cetaceans (the group of dolphins, whales and porpoises), though, and British bottlenose dolphins have been seen killing the much smaller harbour porpoises by constantly ramming and pushing their more diminutive relatives under the water. The bottlenoses are thought to kill the porpoises to decrease competition for the same food supply.

Thanks to their association with coasts and kinship with man, the mating and reproduction systems of common bottlenoses are better known than that of any other cetacean. Gestation is thought to last for twelve months, with British calves being born at any time of year apart from the winter. The calves are born about a metre long and have to learn their mother's unique signature whistle quickly as they will stay close by her side for at least three years. The calves grow quickly on their diet of rich milk, which contains four times the fat of cow's milk, but they do not become fully weaned until they are at least twelve to eighteen months old. Dolphins also have a slow reproductive rate with the average calving interval being between three to five years, but this is offset by the fact that females can live for at least fifty years in the wild.

When a female is in season, the male or males are constantly in attendance and courtship behaviour includes clinging to the female, posing for her, and

Where you see one bottlenose, more will invariably be close by, with this most social of dolphins.

constantly rubbing and nuzzling against her, with the sexual act itself being short-lived but carried out a number of times. In addition to showing their tender side, the males will act aggressively towards females in season and can even be involved in a form of kidnap as they isolate the female and wait for her to become sexually receptive. Males will fight over a receptive female, with the normal technique being a head-butting competition as each male tries to assert his physical dominance. Once the male has mated, he takes no further part in raising the offspring; he is only concerned with his next conquest.

Bottlenoses have a very catholic diet and will prey on a diverse array of coastal fish and invertebrates from the surface to near the bottom, with sea trout, mackerel and mullet often figuring highly on the menu. They are able to work collectively: a pod cooperates to force fish into a shoal and bring them up to the surface to maximise the harvest. In addition to this, they are quite happy to forage individually. During this they have been seen to use a technique called 'fish whacking', which entails striking a fish clear of the water with their tail fluke to stun it before catching and eating it at their leisure.

In places like the North or Irish Seas the visibility quickly declines below the surface, but the bottlenoses are able to rely on a form of echolocation, akin to sonar, to help them find their way around and locate prey at depth. They possess small openings behind the eyes, but most soundwaves are transmitted back through the lower jaw to the inner ear. By making a series of squeaks and clicks with air sacs near their blowhole, the dolphins are able to receive the returning soundwaves and form an 'echoic' image.

In British waters, due to the paucity of sharks, bottlenoses have no natural predators, but, if their habitat deteriorates, it will of course impact on the dolphins too and, while the common bottlenose dolphin is an abundant species worldwide, there are certainly indications that the Moray Firth population is slowly declining due to disturbance, pollution and drowning in fishing nets. However, with current strict protocols on watching dolphins there should be no further disturbance, and it is in the dolphins' flippers as to whether they wish to grace us with their presence or opt for a human-free day from time to time.

Hares

WHEN
Will box most months, but seen more easily while the crops are short in March and April

WHERE
Widespread in lowland grassland, but most abundant in the south and east of England. Havergate Island, Suffolk; Royston, Hertfordshire; Loddington, Leicestershire

14 Boxing hares

There are few wildlife spectacles that are as widely known or celebrated as that of boxing brown hares. In fact, so familiar has this amazing behaviour been throughout the ages, that Lewis Carroll used it as the basis for the 'March Hare' character in his illustrious book *Alice's Adventures in Wonderland*. While the phrase 'mad as a March hare' has also entered common parlance, few people understand the real reason why hares behave in such a seemingly bizarre manner.

The female (Jill) telling the male (Jack) that she isn't ready to mate, in no uncertain terms.

The brown hare is actually by far the better known of our two species of hare, even though it is its cousin, the mountain hare, which is actually native to Britain. Originally emanating from the grassy plains of Asia, the brown hare spread north too late after the last Ice Age to cross the English Channel before mainland Britain was cut off from the Continent by rising sea levels and it is thought the species was introduced by the invading Romans, who were apparently partial to hare for dinner. The hare's actual introduction also pre-dates that of their cousin, the rabbit, by over a thousand years; the rabbit arrived when Britain suffered the humiliation of another invasion, this time by the Normans, in 1066.

The hare's fortunes have always been closely associated with agricultural development. With the historic large-scale clearance of the wildwood to create room for arable crops, more of the brown hare's favoured open habitat was created and now the only areas that remain devoid of this champion nibbler are the uplands and mountains, these niches having possibly been filled by the already established mountain hare.

Unlike rabbits, which go underground to rest, the brown hare is an animal that spends its entire life above ground. It is also largely nocturnal, resting by day and actively feeding at night. However, during the short summer nights they also tend to be active at dawn and dusk. Any hares that are seen running around during the middle of the day are individuals that have usually been disturbed.

When the hares are not seen actively feeding or interacting with their compatriots, they are usually resting in the fields. Most fields have good all-round visibility so the hares are able to keep an eye out for predators, or, when the crop is long enough, they will use it as cover to hide in. If the vegetation is particularly short, or the weather inclement, hares will dig a form, which is a shallow depression in the ground that keeps them out of sight and also helps protect them from the worst of the elements. It is astonishing how the hare becomes virtually invisible in these forms, aided by its camouflage-coloured coat and ability to freeze for hours on end.

Hares spend their whole lives very wary of being eaten and, although often thought of as a solitary animal, they prefer to associate in loose groups when feeding or resting, as many pairs of eyes and ears are much more likely to pick up a fox. Unlike rabbits, hares are unable to bolt down a hole at the nearest sign of danger, so they must rely on fleetness of foot to outpace their enemies. Brown hares have been clocked at an incredible 45 miles per hour full tilt, which is faster than either a greyhound or a mounted racehorse. This is coupled with the fact that they know their home range intimately, and, if chased, will often run along familiar pathways or 'racetracks', enabling them to travel at full speed round various twists and turns without fear of colliding into unknown obstacles. Being light and mobile, hares are also able to change direction quickly and often, a tactic used to outwit their pursuer.

Hare society is not particularly complicated, as the animals do not hold territories, take part in ritualised displays or practise either a pair-bonded or harem system, but there does seem to be a dominance hierarchy with seasoned, experienced animals driving away younger and less assertive individuals. As male hares attempt to mate with as many of the slightly larger and heavier females as possible in a season, it is inevitable that the dominant males will have far more success than their more inexperienced counterparts. While a loose hierarchy predominates with the males, there is no evidence of any ranking system with the females, and they seem to spread themselves evenly out over suitable feeding habitat.

As with so many behavioural spectacles, the reason for the hares' wonderful and mercurial boxing comes down to sex. The breeding season usually begins around mid-February and can carry on through to mid-September in favourable years. The reason why most boxing is seen in March is because this is when

Unlike rabbits, hares spend their entire lives above ground so must tough out the worst the British winter can throw at them.

most females become receptive for the first time; it is also when the crops are still so short that their behaviour can clearly be observed.

As a female comes into 'heat', a number of males can often be seen paying considerable attention to her and relentlessly following her around as she tries her best to ignore them and feed. Quite often 'mate guarding' will occur, whereby a dominant male will constantly station himself next to a female, who is close to being mated, and take it upon himself to chase away any other potential suitors, giving them a sharp bite if they stray too close. As she comes into season, the guarding male begins to pay her more and more attention and is often seen sniffing the ground where she has been as he tries to gauge her readiness to mate.

It was often assumed that boxing hares was a result of rival males fighting each other for mating privileges, but it has now been established that this behaviour is caused by the females fending off the advances of an overly amorous male. If the female becomes irritated with her companion's constant badgering she will turn around in her tracks and give him a box with her forepaws, causing the male to reciprocate, resulting in the so-called 'mad March hare' behaviour.

Eventually the persistent, experienced male will break down her resistance and the female will allow him to mate with her. After all that effort, the actual act of copulation takes place quickly, and often ends with the male making

A female in oestrus is relentlessly pursued and harried by a posse of hopeful suitors.

A hare at full tilt, as it uses its pace and intimate knowledge of its home range to outrun any predators, is an impressive sight.

a little jump away from her. The mating is the sum total of the male's involvement, and both sexes go their separate ways, with the male intent on finding more females with whom to mate.

Hare pregnancy is thought to last for around 41 to 42 days and, unlike rabbit young, which when born underground and are blind and helpless, leverets are born in the open, fully furred and with their eyes open. The usual litter of between one and four needs a minimum of looking after by the female, and, after having been nursed, the leverets move a short distance away from the birth site to separate forms a few metres apart to minimise the risk of the entire litter being uncovered. The mother confines her time with the leverets to just one brief feed every 24 hours, which usually takes place an hour after sunset. At this time, the female returns to the birth site and the young gather round to suckle for less then ten minutes, during which time the female sits upright so she can keep a look out at this vulnerable time.

As smell could play a crucial part in predators tracking down the leverets, when feeding has ended, the leverets will lie on their backs while the mother licks them, and eats any waste faeces or urine the young produce that could potentially give their location away while they lie motionless in their forms. After around a month, the young become independent when the mother does not come back to feed them and they are on their own. Although the first litter will grow quickly and reach adult size and maturity in the same season, it is unusual for them to breed during this first year.

The females become fertile again as soon as they have given birth and may have as many as four litters during the breeding season if the weather is favourable. Mortality among the hares is high and their lifespan is short; a British hare living to the ripe old age of five is considered to have easily beaten the average life expectancy of just over a year. Of course hares make up for this high turnover with prolific breeding, which, in good years, will offset the numbers killed by predators and disease, in addition to those hunted by man, poisoned, hit by traffic or killed during stubble burning.

The number of hares on farms differs widely across Britain, with some areas being devoid of animals, while farms in other locations hold abundant populations. Hares are considered to be most abundant in the arable heartland to the south and east of Britain with a pre-breeding population of around 800,000. From its high point of around four million in the late Victorian era, when small-scale farming and intense predator control was practised, the hare seems to be suffering a long-term decline. Hares have not adapted to living in urban areas or woodland so depend almost entirely on farmers for their continued survival. Often there has been no room for hares as British farming has changed from that of a patchwork quilt landscape to huge sterile fields, which have no cover, and are devoid at certain times of suitable crops for food.

As a result of the increasing popularity of wildlife watching and sympathetic farming techniques, this charismatic animal may, however, have turned the corner. Let's just hope that boxing hares will continue to be an enchanting feature of our arable landscapes for many generations to come.

13 Sea eagles fishing

For anyone lucky enough to travel to Scotland's Western Isles and be treated to views of the proverbial 'flying barn door' that is our white-tailed eagle, it is important to remember that, until recently, this was a spectacle that had been absent from our coastlines for a very long time. White-tailed eagles were brought to extinction in Britain by man at the beginning of the 19th century, but restitution has, in part, been made by a generation of ardent conservationists who rightly thought that the skies were a poorer place without this majestic bird.

With a distance of close to 2.5 metres between each flying wing-tip, the white-tailed eagle has the largest wingspan of any bird in Britain. Even larger than its relative and close neighbour, the golden eagle, the white-tail is a bird of truly leviathan proportions and the fourth-largest bird of prey in the world. When seen in flight, the white-tail is initially more reminiscent of a vulture than an eagle, thanks to its huge deep bill and long neck. The silhouette of its broad wings set off with the deeply fingered primary feathers also contrasts with the vivid bright white tail in the adult birds, the feature that gives it its name.

White-tailed eagles have a sparse distribution from southwest Greenland across to their strongholds in Norway and Russia, with significant concentrations in Poland and Germany. The small but burgeoning British population, all of which originate from Norwegian stock, now stands at just over 30 pairs; there is hope that this upwards trend will continue.

In the 18th century the white-tails were breeding in England, the Isle of Man and their stronghold in Scotland, and it was thought there were at least a hundred eyries. However, the species rapidly declined as a result of persecution by shepherds, gamekeepers, fishery owners and trophy hunters who were after the birds and their eggs. By 1800 the eagles had disappeared from England and were fast declining in Scotland because of a 'talon bounty'. One local man on Skye was said to have killed many eagles – 24 in one year – by shooting those lured down to dead sheep. The last breeding record was on Skye in 1916 and the last British white-tails were shot on the Shetland Islands two years later.

In Scotland the historic and current distribution of white-tailed eagles reflects a strong preference for coastal habitats. An established pair will have a large home range of anywhere between 20 and 40 square miles, of which the key habitat will be a considerable stretch of coastline. White-tails can, and often do, overlap territories with resident golden eagles, but competition between the two species is limited as the latter tend to be birds more of mountain and moorland. On the Isle of Skye, eyries of both species have been recorded as close together as 500 metres; despite this, interactions are rare and their relationship is one of peaceful coexistence or mutual avoidance.

Due to continued protection from conservationists and the boost from tourism it is hoped that this majestic bird of prey is definitely here to stay second time around.

Eagles

WHEN
Can be seen at any time, but
most easily between May to
July when feeding chicks

WHERE
Trotternish Peninsula or
Portree, Isle of Skye; Loch
Frisa on Mull; Gruinard Bay
in Wester Ross

This encapsulates the wonderful array of coastal bird life in Britain, as this common tern is dwarfed by our largest bird of prey.

Each white-tailed eagle territory will usually contain more than one eyrie – eleven were recorded in one instance – and these tend to be positioned on cliff ledges or in the crowns of mature conifers. These eyrie locations may not necessarily be immediately adjacent to the coastline if suitable nesting sites are limited, but adult birds will avoid moving too far from their feeding range.

White-tails are versatile and opportunistic hunters and carrion feeders, and are not averse to using their immense bulk as a bullying tactic to persuade other birds or even otters to relinquish their hard-won gains. Befitting a bird that frequents the coast, however, they eat mostly fish, which, in a similar way to the osprey, they catch by flying low over the water and snatching them in their extended talons from the surface with remarkable precision. Should the fish be slightly deeper down, the eagles have even been recorded occasionally plunging in after quarry. Smaller fish caught are held in one talon and will often be eaten on the wing, while larger fish usually necessitate the use of both sets of talons and a landing back at a favoured perch for the meal to be properly appreciated. Being much less timid then golden eagles, white-tails will even follow fishing boats and feed on the waste that is thrown overboard. As this also happens to be one of the best ways to see the eagles at close hand, a number of enterprising fishermen now make a good living each summer by taking tourists to well-known eyries with a nestful of hungry chicks, in order to 'feed' the adults!

The white-tails will also feed on rabbits, hares and the very occasional lamb which are located from an elevated perch or while in searching flight, and are mostly caught by surprise rather than with any extraordinary agility. While carrion will be included in the eagles' diet at any time, it is particularly important to them during the winter months when other live prey is more difficult to locate. During the breeding season when the adults are busy rearing their young and protecting the territory, the adult eagles will need at least 500 to 600 grams of food a day to sustain them, but they will only need half this amount during winter when they are less active.

White-tailed eagles have a long adolescence and are generally not ready to breed until they are at least five or six years old. They are monogamous birds, forming lifelong bonds; they are also long-lived with an average life expectancy of 21 years, so, if one of the pairing dies, the remaining bird will usually find a new mate. The start of the breeding season is marked with frequent loud calling by the male (and occasionally the female) close to the eyrie in a statement intended to assert their ownership over the territory. Courtship involves high circling of the eyrie location on horizontal wings and will occasionally culminate in a breathtaking aerial display, whereby a pair will lock talons in mid-air and then plunge earthwards in a series of spectacular cartwheels, only to break free just above the ground or water before soaring back up high again.

The nest is built out of twigs and branches and lined with rushes and grasses; as the same site will be used intermittently over many years with incremental additions each season, this means they can often reach a huge size. The female will lay one or two, or very occasionally three, dull-white eggs a few days apart in late March or early April and will begin incubating with the first egg, meaning the hatching is staggered and the chicks will vary in size. While the chicks are mostly tolerant of each other in the nest, when there is competition for food the older chick will dominate and this pecking order has evolved to ensure that at least the eldest chick survives in the lean years. The female does most of the brooding and feeding of the young chicks for the first three weeks while the male's job is to provide enough food for his mate and the chicks. Only once the chicks have acquired the ability to thermoregulate will the female then spend time hunting away from the nest.

The young will grow quickly but are not able to feed themselves until they are at least five to six weeks old. After much frantic wing-flapping to strengthen flight muscles, the young will fledge after ten to eleven weeks but will still be dependent on their parents for a further six weeks while they master the flying and hunting techniques they'll need in their unforgiving surroundings.

Unlike the osprey, which recolonised naturally after being made extinct, the white-tails had to be reintroduced. The programme was started out in 1975 by the Nature Conservancy Council (NCC as it was then) and the Royal Society for the Protection of Birds (RSPB) and spearheaded by far-sighted conservationists such as Roy Dennis. In the first 10 years, 82 young eagles were introduced to Rum in the Inner Hebrides from Norway, and the reward was a first breeding pair in 1985. With additional releases in West Ross in the 1990s the population has slowly continued to climb. The reintroduction has succeeded mainly because habitat loss was never a factor in the extinction and, most crucially, because attitudes towards the birds have changed. Even now, there is no room for complacency as some eagles are still the victims of deliberate persecution and the incidental victims of illegal poisonings aimed at crows and foxes, but most Scottish islanders on places like Skye and Mull are immensely proud of their white-tails. With the added benefit that they are a big tourist attraction, this time it is hoped the white-tailed eagle is most definitely here to stay.

Red deer

WHEN
September and October

WHERE
Highlands and western
Islands of Scotland; Lake
District; Exmoor; New Forest;
selected city parks such as
Richmond Park

12 Rutting red deer

The red deer stag is Britain's largest and most majestic land mammal, and also the beast that features on arguably our most famous wild animal portrait, *Monarch of the Glen*. Painted by Sir Edwin Landseer in 1851, this painting has also become an iconic symbol of Scotland's hills and glens, locations that, to this day, play host each autumn to the red deer enacting one of Britain's most visceral and primeval wildlife spectacles. To see a red deer rut is a powerful experience that dates back to when Neolithic man first occupied Britain and one that must have been enjoyed by spectators through the ages.

This testosterone-charged stag shows other male red deer he means business as he asserts his dominance with deep throaty bellows.

As humans colonised mainland Britain and the ancient forests were cleared, this once-widespread deer was forced to retreat to the most remote areas, with the only remaining English populations of any size now confined to Exmoor, the New Forest, Thetford Forest and Cumbria. Outside these English pockets (and maintained deer parks), the red deer is now mostly confined to the Scottish Highlands and Islands, where a population of over a quarter of a million plays a very important part in both the ecology and economy of Britain's most far-flung corner. With an ancestral distribution that stretches across Europe and beyond, the subspecies of red deer found in Scotland is smaller and has a more diminutive set of antlers than the red deer found on the Continent, a reflection of the harsh and mineral-poor environment in which they live. Nevertheless, a large, mature Celtic stag is a truly impressive sight, tipping the scales at close to 150 kilograms as he prepares for the annual rut.

With the exception of reindeer, in which both male and female grow a full set of antlers, and the Chinese water deer, where neither sex produces any headgear, the red deer is typical of all other deer in that the male is the only sex to develop antlers. In contrast to the horns of sheep and cattle, which are composed of keratin, red stag antlers consist of bone and develop each year from two short outgrowths of the frontal bone called pedicles. Young males grow their first set of antlers in their second year; these are then shed before the beginning of their third year. Each successive year they will then regrow and shed a progressively larger and more complete set of antlers until their seventh year when they are mature and the number of spikes or 'points' levels off, even though their actual length may continue to increase for several more years. In the deer parks, where life is somewhat more comfortable than in the wild, a stag with more than twelve points (six on each antler, including the terminal cup) is a common sight, but a natural 'Royal' or '12 pointer' in the Highlands and Islands is a rare sight for the naturalist or stalker who chances upon one.

When two evenly sized stags clash, it can be both spectacular and brutal. There is much to fight for, as the winner takes all in this most obvious example of 'survival of the fittest'.

Mature stags shed their antlers in February or March and, once the remaining hollow heals over, another stub of bone covered in 'velvet' begins the growth phase again. These new antlers will continue to grow through the spring and summer until the middle of August when a rising level of testosterone in the male triggers a constriction of the blood vessels supplying the velvet, causing this covering to die and rapidly shrivel. The velvet is then usually rubbed off by the stag against trees or the surrounding vegetation to reveal a new set of antlers which are strong, sharp and ready for action. This annual cycle of casting and regrowing a new set of antlers each year is undoubtedly quite an energy demand for the stags, but the advantage for the red deer is that, unlike horns where any damage will be permanent, a major breakage of these antlers will only set them back for a maximum of one breeding season. A new regrown set then means they will be able to compete again on a level playing field the following year.

In addition to the lack of antlers, female red deer, or hinds, are significantly smaller and as much as a third lighter than their male counterparts, this size difference having been thought to have evolved among any deer species where the males have to win and defend a harem. This contrasts with some other deer – namely roe deer and muntjac – where a male need only defend a single female, resulting in sexes of a similar size. In other words, to be a successful red deer stag, size most definitely matters, as the bigger, heavier stags will take the lion's share of the spoils.

The only real time that stags and hinds find themselves together for any length of time is during the autumnal rut as, during the rest of the year, the different sexes keep their distance. Both males and females can be particularly tough to see in the summer in Scotland as disturbance and biting insects often combine to drive them to higher ground during the daylight. Away from the rut, the hinds tend to associate with their immediate female relatives and offspring in a close-knit herd while feeding, while the unrelated stags tend to be more widely dispersed in loose aggregations as they concentrate on putting on weight after the rigours of the winter. These groups of hinds have well-defined home ranges that are rarely larger than three-quarters of a square mile, in contrast to the more mobile stags, which have separate summer and winter ranges, in addition to defined rutting areas.

As well as growing a new set of antlers and bulking up, the stags' testes also triple in size prior to the rut. The increased level of testosterone at this time leads to the development of more powerful neck muscles and a mane and changes their behaviour; they become increasingly agitated as they thrash their antlers against the vegetation to ensure they are in tiptop condition. Late September eventually sees the male assemblages fragment as they move to the rutting grounds.

Once at the rut, the large competing males will spend long periods roaring at other stags as they try both to assert their dominance and to gauge the fitness levels of their rivals. These guttural bellowing calls can be made for as

The quintessential image of the *Monarch of the Glen* ready for battle against a backdrop of russet autumnal tones.

long as 20 minutes at a time, with the noisiest stags often being the most successful; incumbent harem holders always roar more frequently than their hind-less challengers. As the risks associated with fighting can be quite high – as many as 20 per cent of stags can be at least temporarily injured and a small proportion permanently disabled or even killed – many of the smaller challengers will be seen off at this stage rather than risk taking on a mature and experienced stag. Fights only take place when a serious challenger not intimidated by the resident male's bellowing decides to attempt a coup d'état by advancing towards his adversary and initiating a 'parallel walk' as each sizes up the opposition. After strutting for a while, one of the stags will decide to cut straight to the chase and lower his head, which is seen as a cue for the rival to do the same; they then charge together and lock antlers.

Fights can last from a few seconds to over five minutes as each stag tries to lunge through his rival's guard by twisting, turning and pushing. The more experienced male will often use gravity to gain the initiative by driving his rival downhill and away from his harem. It will need a very powerful and determined challenger to drive a resident male away from his hinds but, with the winner mating ten to fifteen hinds or more, the stakes could not be higher. At locations such as the Isle of Rum, where the red deer have been studied for a number of years, experts like Tim Clutton-Brock reckon that as many as 80 per cent of the calves are fathered by just 20 per cent of the stags, with the most successful males consistently being above average size and weight from youth all the way through to adulthood. Interestingly, while the actual presence of antlers is an essential prerequisite to repel other males and defend their hinds, the size of the antlers is not considered the most important factor in comparison to sheer bulk.

The hinds rarely come into oestrus for longer than 24 hours and during this small window of opportunity they will signal their readiness to mate by displaying an intense interest in their harem holder. In Scotland the mating peak generally falls in the second half of October, with most copulation taking place in the early morning or evening as the stag mounts each receptive hind several times. The mating and constant defending of their harems is so time-consuming for males that they rarely eat during the rut and can lose as much as 20 per cent of their body weight before they drift away exhausted to their winter quarters. Under such intense annual competition, even the largest stags are rarely at the top for many seasons and a male that lives longer than twelve years is considered a veteran.

Calves conceived by these autumn matings are born from mid-May to early June and initially kept in long vegetation, where they are visited by the mother only two or three times a day. At two to three weeks old the mother takes her unsteady offspring to join her usual group where, depending on its sex, the youngster will either spend her life in her adopted clan, or bide his time dreaming of glory on the battlefield.

Red deer can be a threat to the regeneration of trees and shrubs so controlling deer numbers while achieving a sensible balance with conservation of the species is a major challenge. Apart from the huntsman's bullet, however, the only real threat to the red deer – bar those populations on isolated or remote Scottish islands – comes from hybridising with the introduced sika deer.

Geese

11 Flocking pink-footed geese

The pink-footed goose cannot, in looks, compare to the elegance of the mute swan, the dazzling plumage of the kingfisher or dashing nature of the peregrine. But the goose makes up for its unremarkable looks and modest behaviour by ganging together in huge flocks, making for a spectacle – in the truest sense of the word – where the whole is indeed greater than the sum of its parts. Not only does Britain play host to the vast majority of the world's pink-footed population each winter, but this goose is a conservation success story. Since 1950, the numbers over-wintering here have increased eightfold to approximately 241,000 birds.

The pink-foot is a medium-sized goose that measures 60 to 75 cm in length, putting its dimensions somewhere between that of the larger greylag (with which it often associates) and the other common estuarine winter goose, the 'mallard-sized' brent goose. The pink-foot's name obviously derives from its pink feet, which complement a short pink bill, but, at a distance, the most distinguishing feature is its much darker chocolate-coloured head that stands out against the shades of grey, buff and white that make up the body plumage.

The 'pink-foot' has the smallest breeding range of any of the grey geese, choosing to rear its young in a narrow band from eastern Greenland to central Iceland and Svalbard. As the species migrates to northern Europe, the entire populations from Iceland and Greenland pass the period between mid-October and mid-April at a few prime British sites, while the much smaller Svalbard breeding population over-winters in the Netherlands and western Denmark.

The only time that this sociable and gregarious goose does not gather in huge flocks is during the breeding season. Adult pink-feet will generally pair up in their second or third year, with the bond often lasting for life. Their nesting locations, in Iceland for example, will be in inaccessible river gorges or on cliffs next to glaciers, where their young will hopefully be safe from ground predators such as the arctic fox. The clutch is laid in mid-May and usually consists of between three and six eggs, which will be whittled down to two or three goslings by the time the juveniles are ready to leave the nest. On their breeding grounds, terrestrial and aquatic tundra plants form the mainstay of their diet, with both parents helping to tend the young as they form a familial bond that will stay intact throughout the migration and winter, and will only begin to break up as the adults prepare to return for the following breeding season.

When the young goslings are 10 to 20 days old, the family party leaves the nest site and tends to congregate with other groups while the adults' flight

Skeins of 'pinks' heading for the roost after a day spent helping themselves to the remnants of the sugar beet crop is undoubtedly one of our most evocative winter wildlife images.

feathers moult in preparation for the journey south to Britain. This 'safety in numbers' technique affords some protection during this flightless and vulnerable time, but they are also able to cover large distances on foot to avoid areas where there may be large numbers of predators. After a period of about 25 days, the adults' new feathers are ready for action and the young can also fly, and at the beginning of October there is a mass emigration across the Atlantic.

The historical and traditional areas for pink-feet to spend the winter are the Grampian and Aberdeenshire regions of northeast Scotland, Tayside and Lothian, Dumfriesshire, Lancashire and Norfolk. A long-term ringing programme has revealed that newly arrived geese will firstly spend the early winter in the Grampian or at reserves along the coast such as the Loch of Strathbeg in Aberdeenshire, but will rapidly move further south into England and Lancashire with many finally arriving in Norfolk by mid-winter.

The sight of flocks, or skeins, of geese flying to and from their roosts is one of the greatest winter highlights and, at its best, is even representative of the archetypal image of wilderness Britain. In Norfolk, which now has the largest population of pink-feet – between 70,000 and 110,000 at various stages during the winter – the geese tend to be creatures of habit with well-known roosting and feeding sites and clearly prescribed times for moving between these spots. Most of the geese spend the night out on the estuary where there is little or no danger from predators but, as dawn breaks, they fly inland in a series of waves or squadrons that must bring back memories of the Battle of Britain! These massive flocks can, from a distance, look like a series of fast-moving clouds, but, as they move on to and over the land, they often assume a series of V formations in the sky, with smaller Vs often being slotted inside larger versions. The function of this V is to confer an aerodynamic advantage to each individual as the downbeat of each wing produces a corresponding swirl of rising air, which the bird in the slipstream can then tap into in order to give it extra additional lift and forwards speed. The trailblazing goose will be an adult with an extensive knowledge of the feeding and roosting areas and the leading bird will often change position periodically to give each a rest. In practice, this precise V formation will often morph or mutate into a shapeless formation, resembling a child's scribbles across the sky.

These skeins are often accompanied by a huge amount of noise as the geese call excitedly to one another; from a distance their calls can sound like a pack of dogs. When closer, the incessant call sounds like a musical 'wink wink' that is

While having pink feet, the most distinguishing feature of this smart goose at a distance is its chocolate head ... time for a name change perhaps?

considered higher pitched and less harsh than the other geese. Feeding areas are rarely more than 18 miles from the roosting site, and often will be considerably closer, with returning geese knowing their home range intimately When the suitable feeding area is located, the geese will then bow their wings as they lower their altitude, and, when they are 20 or 30 metres above the field, they frequently slideslip and tumble quickly down to the ground as they spill the air from their wings in a process known as 'wiffling'. Even while on the ground, the noise levels are high as the family parties maintain contact; the flock's chattering also serves to suggest the flock is not under threat and it is safe to begin eating.

Being herbivores, pink-feet feed on a variety of vegetable matter including grain, winter cereals, potatoes and root crops. The exponential increase of this species in Norfolk – from none in 1975 to the current staggering total of 30–40 per cent of the world's population – is, however, mostly due to the increased production of sugar beet. The post-harvested remains of this root vegetable provide a carbohydrate-rich meal for the geese, and, because the beet fields are larger since the hedges have been grubbed up, these super-wary birds now have added security thanks to clearer all-round vision. In addition, while the geese continue to be shot, a series of Wild Bird Directives in the 1970s banned their sale, meaning that wildfowlers now have no commercial market in which to sell the geese beyond those for their own consumption.

Flocks in certain sugar beet fields can often number close to 10,000 so there are always plenty of pairs of eyes to keep a constant look out for any potential threats, and many of the geese can put their heads down and switch to grazing mode. These large feeding flocks are always a flurry of activity, with ganders chasing one another with necks lowered and wings spread as they defend a small feeding patch for their family. At the good feeding sites, groups will constantly be flying in and jostling for the best feeding spots. Though the geese undoubtedly do a degree of economic damage when they graze on winter crops and grass leys, by feeding on the waste leaves, stalks and the root crowns of the sugar beet, they may reduce transmission of crop diseases from one year to the next.

Constant disturbance can be one of the major problems for the pink-feet as they can waste valuable energy being pushed between various feeding areas. Where interruptions are unacceptably high, even good feeding areas will often be abandoned. During moonlit nights the geese may avail themselves of the extra light and spend most of the night feeding in the fields, but usually dusk signals the time when the feeding birds take to the air to return to the sanctuary of their coastal roosting sites until the following dawn. With a setting sun in the background, dusk is also often the best time to appreciate the constantly shifting patterns of the skeins as they fly overhead. This spectacle has been a source of inspiration for a number of leading luminaries, such as wildlife artist and conservationist Sir Peter Scott, who used flying geese as one of the most common themes in his paintings.

10 Barn owls hunting

Unfortunately, the most common sighting of this stunning bird is now that of an occasional ghostly glimpse in the car headlights. Outside a few rural strongholds, our much-loved and familiar barn owl has declined dramatically over the last 80 years as it has struggled to adapt to the modern way of life in Britain's countryside. Despite the well-charted declines in Britain and Europe, however, the barn owl is one of the most widespread bird species in the world, with 35 different recorded sub-species, which are found on every continent except Antarctica. The only regions where barn owls are completely absent tend to be the northerly latitudes or mountainous areas, where the winters are simply too harsh, or the habitats of tropical forests and deserts, which are unsuitable.

The barn owl is one of Britain's most easily identified birds, with its relatively long wings, which enable it to fly very slowly or even hover while hunting, and long legs, which are used for catching its prey in the long vegetation. However, surely the owl's most distinctive feature is its beautiful pale and buff plumage, from which the bird takes one of its colloquial names, the 'white owl'. Another commonly used local name is that of 'screech owl', which comes from the long, drawn-out screaming call made in flight. The barn owl has, in fact, a wide repertoire of hisses, shrieks and snoring sounds, the latter being commonly heard during the breeding season and made by the female and young; it has a remarkable resemblance to the sound of Dr Who's Tardis! When heard in the dead of night, these calls have an unworldly and eerie feel about them and it is no surprise that the naturalist Gilbert White, in his famous book *The Natural History of Selbourne*, wrote that whole villages were 'up in arms on such an occasion, imagining the churchyard to be full of goblins and spectres'.

In terms of size, there is little difference between the sexes of the barn owl, with both measuring a touch longer than a standard ruler at between 33 and 35 centimetres. The female is generally a little heavier than the male, reaching a peak of 400 grams during the breeding season before losing weight again during the winter. The males can be identified by their paler upper parts, which are also less barred than the female, and virtually snow-white underparts, which contrast with the flecked breast, belly and flank feathers of the female. British barn owls are also one of the palest sub-species; the Continental forms are much darker on the back and honey-coloured on the breast.

The barn owl's preferred habitat is open or lightly wooded country and it has historically benefited from man's clearance of the land. In terms of nesting sites, the barn owl is more than happy to lodge alongside humans, with barns (from which it gets its name) and other disused farm buildings being very desirable locations. Alternatively, barn owls will nest in more natural sites such

Britain's countryside would be much poorer without this enigmatic beauty.

as an old tree hole or cliff ledge where permitting. Interestingly, there does seem to be a bias towards barn owls using more natural nesting sites in the drier east of Britain; those in the wetter west seem to prefer the more sheltered accommodation of farm buildings. Unlike fiercely territorial tawny owls, a number of barn owls may nest in close proximity where prey is abundant.

The barn owl plies its trade as a small mammal hunter with a range of perfectly evolved senses that enable it to pinpoint and kill its prey during the darkest of nights. Its hearing is incredibly sensitive, with any noise being gathered by its heart-shaped face, which effectively consists of two oval and concave discs of short, stiff feathers that funnel the soundwaves down to the ear openings and the eardrum. These two openings are placed asymmetrically so sound can be pinpointed with devastating accuracy even in the pitch-black.

Their eyes also operate well in very low light conditions, as an enlarged cornea and lens focus what little light is present on to a retina that is packed with a high density of light-sensitive rod cells. The barn owl's eyes also function effectively when hunting in broad daylight, but the trade-off between having such incredible sight at low light is that they may well perceive colour more poorly than daytime birds. In common with virtually all predators, the barn owl's eyes also face forwards, giving it excellent distance perception.

Their hunting technique consists of slowly and deliberately quartering the

A mouse meets its maker in the talons of this ruthlessly efficient predator.

Barn owls

WHEN
Anytime, with summer evenings being particularly productive

WHERE
Widely distributed in lowland Britain; Norfolk, Devon and Yorkshire are particularly rich counties

grasslands on buoyant wings at a height of between two and four metres, stopping frequently to hover in mid-air when they hear anything of interest or spot movement. Their feathers are dampened to vastly reduce flight noise, with specialised hairlike extensions that extend from the feather barbules (that fringe the barbs of the feather) and give the owl's plumage the familiar soft, velvety feel. This enables the barn owl to make a virtually silent approach when hunting so that the prey item is unaware of its presence, and also so that there is less noise preventing the owl from accurately calculating exactly where to drop.

Although they will hunt from perches such as fence posts, pouncing usually takes place from the air, with the owl frequently hovering before dropping quickly into the vegetation with legs extended forwards, talons spread and wings held aloft above the body. Even while diving, the owl is capable of making tiny adjustments and, upon snaring the prey, the wings fold over the doomed rodent while it is dispatched. Occasionally, the birds can even be seen chasing their prey on foot if the prey escapes the aerial assault.

Unlike birds of prey such as sparrowhawks or peregrine falcons, which frequently dismember their prey at the site of capture, barn owls usually swallow their food whole immediately unless the food is to be carried back to feed a hungry brood. As the owls' stomach acids are not able to digest fur and bones, these remains are retained in the foregut and eventually become compressed and regurgitated as dark cylindrical four to six centimetre-long pellets. These pellets may well be the first evidence of a resident barn owl and are usually found below roosting sites or favourite perching locations. When teased apart, complete skulls or identifiable jaw-bones are frequently found in among the other assorted bones and fur of the pellet, and these can be a mine of information on the barn owl's prey items.

While the barn owl's diet will include brown rats, common shrews, wood mice, beetles and even occasionally birds, without doubt the most commonly caught and important food item is the short-tailed field vole. This vole is thought to be Britain's most abundant mammal, with populations reaching 75 million at their peak, and can be found across a wide variety of habitats from rough grassland to conifer plantation. The vole also tends to have a cyclical abundance with boom and bust years, which directly impacts upon the breeding success and survival rates of the barn owls themselves.

The only weather that prevents barn owls from hunting is rain, as the owls seem to detest getting their plumage wet, but deep snow and prolonged icy periods will also prevent access to their prey, and mortality can be high during particularly cold winters. For birds that see the following spring, however, the barn owl breeding season begins very early in the year with regular visits to the nesting site as the male proclaims ownership of the site to his mate. Males without mates will often stock up a larder of food items to impress any visiting females with their catching prowess and the quality of their territory. This supply will also help to improve the female's condition in preparation for the arduous processes of egg laying and incubation.

The laying date depends on the abundance of voles, with clutches of four to six laid on alternate days in among a rudimentary nest of a pile of old pellets. Incubation begins with the first egg meaning that, as the eggs hatch asynchronously, a dominance hierarchy develops. Only the older chicks will survive in years with poor vole numbers, and, with reports of the larger occasionally resorting to eating their younger siblings, the successful barn owl fledgling is a living embodiment of the phrase 'survival of the fittest'! After two to three weeks, the chicks are able to swallow food whole and visits by the parents are restricted to passing over food. While the young will reach their final weight by 35 days, fledging does not occur until wing growth is completed at between 60 to 65 days. The young are then fed for up to a month before they disperse locally.

Most barn owls will start breeding in their first year of life and, once they have settled at a site, they are remarkably sedentary. Barn owls have few natural predators apart from the occasional opportunistic goshawk or buzzard, so the main causes of death are starvation of inexperienced birds, accidents on the roads or, bizarrely, drowning in water troughs when trying to take a drink.

In most of Europe barn owls have suffered from the large-scale intensification of farmland by increased mechanisation, removal of prey-rich pastures, grubbing up of hedgerows, application of fertilisers and pesticides and a reduction in nest sites thanks in part to barn conversions. From an estimated population of around 12,000 pairs in 1934, they now number little more than 2,000 pairs. However, with the successful advent of nest boxes and more sympathetic management, the barn owl may well be about to turn the corner.

9 Basking sharks feeding

Anaesthetised with a diet of exotic wildlife programmes filmed in far-flung locations, it is easy for us to believe that bigger and better wildlife spectacles only exist elsewhere on the planet. But there is one truly world-class wild experience that can be seen every summer along our own Atlantic coastal waters, and a close encounter with the world's second-largest fish, the basking shark, is one of those life-affirming experiences you simply can't afford to miss.

Second in size only to the whale shark, the basking shark is the true leviathan of our British waters. Reaching lengths of eleven metres and weighing up to seven tonnes, this is a very impressive and truly unmistakable fish. This can lead to errors in identification, however, when sensationalist fans of the *Jaws* movies occasionally misidentify basking sharks as great whites. In contrast to their fearsome cousins, basking sharks are considered gentle ocean giants, but with the obvious caveat that any animal weighing close to the same amount as a double-decker bus should be treated with total respect at close quarters!

The basking shark's name arises from the fact that, when encountered during the summer months, it always seems to be basking on or very close to the surface, particularly when the weather is calm and sunny. In addition to this habit and to its enormous size, a basking shark can also be recognised by its five long gill slits, which almost encircle the head and are specially adapted for filtering enormous quantities of plankton from the water.

The basking shark is a fish with a huge distribution; it can be encountered in cold to temperate waters anywhere from the Atlantic to Pacific and Indian Oceans. Closer to home, it is most commonly seen between the months of May and October around the Cornish and Devon coastlines and up to the seas around the Isle of Man and western Scotland. Basking sharks are very rarely seen during the winter months, which led, initially, to speculation that they spent this period in hibernation, but recent research work suggests that, though they stay in the vicinity of our waters, they venture out to feed in the much deeper waters off the continental shelf to the west of Britain.

The basking shark attains its gargantuan size on nothing more than a diet of microscopic zooplankton. Zooplankton is actually a term for a 'marine soup' of copepods, larvae, eggs and oceanic shrimps, which are a vital link in the marine food web and the principal fodder for a vast array of marine animals from the smallest fish to the largest whales. The basking shark's harvesting technique of these minute organisms is very simple: it cruises along at very leisurely speeds of around two miles per hour with its mouth agape; it then sieves the zooplankton out of the phenomenal equivalent of a fifty-metre Olympic-sized swimming pool of water every hour. Unlike the other two species of filter-

feeding shark, the whale and megamouth sharks, however, the basking shark's feeding process is carried out in a totally passive manner with no active sucking in of water at all. While the basking shark does possess a set of small teeth, they play no part in feeding; the zooplankton are extracted by specialised gill-rakers, which are comblike structures attached to the gills that trap the food before the water is expelled through the shark's long, modified gill slits.

While the shark is actively feeding with an open mouth, the gullet is firmly closed to prevent water rushing into the shark's stomach. As zooplankton-free water is expelled, the gills extract dissolved oxygen as the water flows over them before passing out through the gill slits. Because of the sheer quantity of water that is filtered and the food extracted by this sieving technique, it has been estimated that, in good feeding areas, basking sharks may well have up to half a tonne of food in their stomachs at any one time, which is certainly one super-sized meal.

This constant search for food means that basking sharks' movements are invariably dictated by the location of plankton hot spots, which often tend to

Basking sharks

WHEN
Mid-May to September

WHERE
Mostly off the coast of
Cornwall, Devon, the Isle
of Man and Dumfries &
Galloway

Why these sharks are
called 'megamouths'
is not hard to fathom
as they trawl the
British waters
for microscopic
plankton.

A basking shark typically swims with its mouth open for anywhere between 30 and 60 seconds before closing its mouth and swallowing a few times, in a manner that has been likened to Kermit the Frog!

form at 'current fronts', where two water masses of different temperatures meet. It is known that the sharks are able to locate these areas of abundant food across ranges of over 300 miles, but it is still uncertain if they have the ability to sense these blooms over long distances, or if they have merely learnt where and when to find these best feeding spots.

In summer the zooplankton blooms are most abundant close to the surface and the first sign of a shark's presence is usually when the dorsal fin, the upper lobe of the tail and occasionally the shark's bulbous snout break the water surface. The shape and condition of the dorsal fin can vary enormously with each basking shark and can be used to identify certain individuals. This, in turn, is helping us to understand the movements and ultimately the conservation priorities of these enigmatic creatures more clearly. Basking sharks seem to be quite social animals, and, in areas of high plankton concentration, can range in number from singletons to small groups, or, in a few cases, schools of hundreds of sharks at particularly rich feeding locations. Interestingly, in Britain, the vast majority of observed sharks seem to be adult females, often outnumbering the males by 40 to 1, and it is thought that, for large parts of their life cycle, each sex seems to prefer the company of their own kind.

Surprisingly little is known about basking shark reproduction, but it is thought that males mature at twelve to sixteen years or around five metres in length, while females are not ready to mate until they are at least twenty years old or an impressive eight metres long. Basking shark courtship does occur in British waters and can involve behaviour such as nose-to-tail following, parallel swimming and even breaching, where the shark is able briefly to leap out of the water. This breaching was initially thought to be carried out to remove parasites such as lampreys or remoras, but it is now thought that the behaviour stems from the time-honoured mating game, with males displaying their prowess to one another or with females displaying their readiness to mate. Breaching basking sharks can often form the crescendo of an already impressive performance and these individuals may well exit the water at least three times in quick succession, while other whale sharks close by carry on swimming or feeding in a totally unconcerned manner!

While courtship has been regularly witnessed, mating has been seen only once in the western Atlantic off the coast of Nova Scotia, and it seems to involve the male using his small teeth to hold on to the female's dorsal fin while the deed takes place. The length of pregnancy is shrouded in mystery, and has been estimated to last anywhere between a year and twenty months; the act of giving birth is also little understood, and has also only been recorded once, by a Norwegian fisherman who witnessed the birth of six live young of around one-and-a-half metres in length.

Basking sharks are thought to live up to at least 50 years old, but, because of their low reproductive rate and slow maturity, they are very vulnerable to hunting. Various parts of their bodies are considered valuable and, consequently, they have disappeared from large areas of their former range and are now listed globally as Vulnerable on the International Union for the Conservation of Nature's (IUCN) red-data list. The shark has a huge liver, which accounts for 25 per cent of the animal's weight and provides a 2-tonne shark with near-perfect neutral buoyancy in the water, but this organ is also highly sought after as it contains large quantities of oil that can be used in the tanning process and for machine lubrication. Additionally, the fins can be sold for ridiculously high prices to the Asian markets as a soup ingredient.

In British waters, the only problems that basking sharks are likely to encounter are disturbance by boats and jet skis, and the occasional overkeen naturalist. There is a strict code of conduct in place to ensure that they are left to graze in peace, which includes a ban on running outboard motors within 100 metres of the sharks to avoid them being hit by propellers, and a rule that snorkellers or divers keep at least 4 metres away from sharks at all times. It is worth noting that the only human deaths attributed to basking sharks occurred in the Firth of Clyde before World War II, when a breaching shark, accidentally hit and managed to capsize a small boat that had drifted too close to the sharks, drowning three of its occupants.

Swimming next to these huge masters of their own environment is a privilege and a tremendously powerful experience. And, if the sharks continue to be protected, conserved and treated with due reverence, there is no reason why it cannot be enjoyed by generations of British wildlife watchers in the future.

8 Dancing adders

There is something both undeniably thrilling and slightly scary about seeing a snake. With Britain's latitude, and snakes primarily being species that like it warm, however, we only have three species to call our own; of these, the adder – as our only venomous snake – is by far the most celebrated and feared. Timid by nature and rich in folklore – though hardly deserving of its villainous reputation – the adder is hard to track down, but, as any self-respecting naturalist will tell you, if you are willing to make the effort, then, with luck, you will be justly rewarded.

The adder is also called the viper, which is the family of snakes to which the species belongs. The vipers are considered the most evolutionarily advanced and sophisticated of all the snakes and include in their ranks some of the world's most poisonous representatives, such as the puff adder, gaboon viper and fer de lance. While the adder is nowhere near as dangerous as its tropical cousins, nevertheless it should be treated with caution and respect.

The origin of the name 'adder' is believed to be the Anglo Saxon word *naedre*, meaning 'a creeping thing'. In Britain the adder can be found with a patchy distribution in England, Scotland and Wales, but is absent from Ireland and islands like Orkney, Shetland, the Outer Hebrides and the Isle of Man. While it can be found in sunny and undisturbed habitats such as forest edges, old quarries, sand dunes and moorland, it is most abundant on sandy heaths in southern England. Despite being cold-blooded, the adder is tolerant of low temperatures and has successfully colonised huge areas of temperate Europe and Asia; it is the only snake species to be found inside the Arctic Circle in Finland!

The adder can be told apart from our two other British serpents, the grass snake and smooth snake, by the bold zigzag pattern that extends along the entire length of its back. The adder is also most heavily built and is the only British snake to have an elliptical pupil as seen in a cat's eye. In addition, it is one of the very few snakes where the sexes can be differentiated by colour. The males tend to look brighter, with their creamy-yellow or near-white background colour making the dark dorsal pattern and the markings along the flanks stand out much more clearly than the dull reddish-brown or yellow-brown background colouration of the female. Both sexes have a very distinctive eye colour, which perfectly matches the colour of autumnal copper beech leaves.

Interestingly, black adders, from which Rowan Atkinson's famous comedy character took his name, are not as uncommon as once thought and certain locations have unusually high numbers. The black adders may well be able to warm up more quickly than their normally coloured counterparts, meaning they will still be able to hunt on cool days. But this benefit may be balanced by the

Exactly what is it that's not to be liked about adders? They are strikingly beautiful, endangered, fascinating and misunderstood.

Adders

WHEN
Early mornings in April and
May are the best times to see
basking adders

WHERE
Heathlands of Exmoor, the
New Forest and the Gower
Peninsula among many
other scattered locations

fact that they are not as well camouflaged and so can be spotted more easily by natural predators such as buzzards, crows and ravens, or the occasional human.

Adders rarely exceed 60 centimetres in length, with the males generally marginally smaller and shorter, with a proportionately longer tail. Juvenile adders grow quickly in their first five years after which they will increase in size much more slowly. It is thought that that in the wild they may well live to at least 15 years old, with 25 years having been reported by some herpetologists.

Being a cold-blooded creature, the adder relies on the warmth of the sun to perform all its daily activities. The snake derives the necessary energy by basking in warm and sunny south-facing spots out of the wind, and maximises the amount of heat absorbed by spreading out its ribs, which flattens the snake and ensures that the largest possible surface area becomes exposed either to the sun or to the warming ground underneath. As the adder is incapable of generating heat, it has to hibernate during the autumn and winter to avoid the freezing weather. The snakes generally become more lethargic when the days become shorter and colder, and finally enter their hibernacula at the end of September or the beginning of October. Hibernation spots can vary from earth banks to rodent burrows, but, crucially, must be both dry and frost free. As good hibernacula can sometimes be hard to find, in tried and tested spots, the adders may well congregate communally.

Adult male adders are always the first to emerge around the end of March, and will spend the first two to three weeks basking close to the hibernacula as they use the sun to speed up the development of their sperm. They can be very vulnerable during this period of little mobility, and stories abound of gamekeepers dispatching several basking adders with a single shotgun blast in early spring. The females and juveniles generally emerge between two and four weeks later and often use different basking spots away from the hibernacula.

The snakes' sluggish behaviour changes radically once they slough their skins in April. By this stage the males will begin travelling up to 200 metres a day actively looking for breeding partners; this search is made easier as the females release a scent from a pair of glands at the base of their tails that serves to attract any male in the vicinity. Once a male encounters a prospective mate, he will constantly flick his tongue along her back and sides and move alongside her in a rather jerky fashion. Both snakes will continue to tongue flick and body quiver for a few minutes before the male eventually aligns his body to enable copulation to take place. The pair may remain coupled for over an hour, while the male transfers his sperm and then inserts a plug in an attempt to prevent the female mating with any other males.

Should another male turn up while courting is in progress, a battle will ensue, which will often terminate in the famous 'adder dance'. This involves the rival males rising up with their bodies intertwined as they attempt to wrestle each other to the ground. The fight often involves much swaying from side to side as each tries to push home his advantage until one of the contestants gives in and flees for cover. Interestingly, at no point do the rival snakes try and bite each other. The winner of these fights is nearly always the larger specimen who is then free to return to his courtship with the female.

The energetic demands placed upon the females, from a gestation period lasting for between three and four months out of the total six-month period spent above ground, mean that it is thought that they only breed every other year. The litter of between five and fourteen young are born live at the end of August and will stay with their mother for a few days after being born. Although the 14-18 centimetre-long young do not usually feed immediately – they are able to nourish themselves on their embryonic yolk supply – even at this early stage they are still capable of delivering a venomous bite.

Adders are masters of surviving for long periods with no food without any ill effects or loss of weight. The reserves of fat laid down during the good times of summer will see them through hibernation and courtship when they are more occupied with mating than finding a square meal. Immediately after mating they will move to the summer feeding areas, which can be over a mile away and tend to be wetter, more low-lying and with an abundant food supply. This then becomes home for the next four months as they fatten up before they slowly make their way back to the same hibernacula.

The adder's diet primarily consists of mice, voles, shrews and lizards, but they are not averse to taking fledgling birds, and one adder was once even seen trying to consume the chick of a merlin! The snake uses a combination of the 'sit and wait' strategy, whereby anything that bumbles across its path may well be opportunistically dispatched, with active hunting. When actively searching for prey, the adder will move carefully through the undergrowth until it either catches sight, or picks up the scent, of a potential meal. It then slowly stalks the animal until it is very close, draws back its head in an S-shaped loop, and strikes with its hinged fangs, before releasing a fraction of a second later. The wounded animal often dives for cover in a bid to escape, but within a couple of minutes the venom takes hold and the adder then follows the trail of the dead or dying animal by smell. Not every strike will be successful and sometimes the adder has to stalk an animal several times or follow a rodent down into its burrow where a kill can be more easily made. The prey is swallowed whole, head first and as quickly as possible, as the procedure can be a vulnerable time for the snake if it has caught something large like a vole. After a big meal the adder will then rest up for three or four days while digesting.

The adder population has undoubtedly declined due to mass loss of habitat and only really flourishes where disturbance is minimal. Not by nature an aggressive animal, given a choice the adder will always retreat. Of the 90 people bitten each year, only a small minority are bitten accidentally; the vast majority of bites happen when humans try to catch or kill the snake. With no attributed adder death for 20 years, however, sensible precautions can ensure that the naturalist will take away nothing but photos and great memories.

Stepping on an adder will happen incredibly rarely as, given the chance, these alert and shy creatures will always slip into the undergrowth rather than enter into a confrontation.

7 Bluebell displays

In 2002, the bluebell was given the heady accolade of being Britain's national flower and, when growing en masse in a spring woodland, it creates a dazzling display that surely ranks as our greatest botanical spectacle. This annual floral extravaganza is also a peculiarly British phenomenon, with an incredible 50 per cent of the entire world bluebell population being confined to our shores; the remaining bluebells are meagrely sprinkled along the Atlantic fringe in a wide arc from Spain to Holland.

Britain's bluebells have been celebrated by generations of poets such as John Clare and Gerard Manley Hopkins. It is difficult not to feel poetic when either striding through a host of nodding bluebells – which can give the impression of wading ankle-deep through swirling water – or when viewing a bluebell bank from a distance, where the massed colour can seem like smoke drifting across the woodland floor.

Although bluebells are considered to be, first and foremost, a woodland species – around 70 per cent of the displays are among trees – they do not necessarily need woodlands to thrive. In the wetter west and north of Britain, for example, it is thought that humidity and continuity of habitat are the vital ingredients for successful growth, resulting in bluebells growing in profusion along lowland hedge banks or upland pastures, which may well have been devoid of trees for hundreds of years. In certain Atlantic-facing locations, bluebells will also flourish on cliff tops that may never have seen trees. One of the very few habitats where bluebells really struggle is under conifer plantations, as the shed pine needles result in conditions being to far too acidic for the plant's liking.

Despite being mostly found under woodland, bluebells are intolerant of deep shade, and they solve this dilemma by ensuring their main period of growth above ground occurs before the trees' leaves begin to properly unfurl and the canopy casts the woodland floor into shadow. It is thought that, during the key growing period from early April to mid-June, as long as at least 10 per cent of the light reaches the forest floor, this is sufficient to enable growth and eventual flowering. Bluebells can be found on a great range of soils from that of clay to chalk, but grow best in earth that is well-drained and nutrient poor. For the bluebells, nutrient uptake in the soil is aided by association with a fungus around the plant's roots (known as a mycorrhizal association), which helps it source a range of minerals, particularly phosphorus.

The bluebell shoot is shaped like a spear and is specially strengthened to ensure it is able to push its way up through the mass of decomposing leaf litter in January. Once the tip breaks the surface, growth stops temporarily until March when the growth of the narrow straplike leaves then proceeds apace.

A bluebell woodland spectacle is a phenomenon truly unique to Britain.

Bluebell woods

WHEN
Dates may vary according to the location, but generally mid-April to mid-May

WHERE
Many woodlands including West Woods, near Marlborough, Wiltshire (Forestry Commission); Bates Green Farm, Arlington, East Sussex; Carstramon Woods, near Gatehouse of Fleet, Dumfries & Galloway

It's not difficult to see why generations of poets and writers have felt compelled to wax lyrical about bluebell woodlands in spring. When seen at their best, allusions to smoke and waves are easily made.

The leaf formation is quickly followed by the production of the flower stalk, with eventual flowering occurring anytime between late April and June depending on a combination of the plant's geographical location and the local climatic conditions.

Although the colour of the flowers can be quite variable, with completely natural white and pink variants regularly cropping up, by far the most dominant colour is a deep violet or cobalt blue, a surprisingly rare colour in nature, and the feature from which the bluebells obviously derive their name.

The flowers are shaped in a narrow straight-sided bell, with six peeled-back lobes around the fringes that are set off in the centre by an array of creamy coloured pollen-bearing anthers. The number of bells can vary between five and sixteen and they are positioned in a loose cluster at the top of the almost entirely one-sided flowering stem, which causes the bluebell to nod or droop at its tip. When looking at massed ranks of flowering bluebells, the visual feast is enhanced by the fact that all the bluebells often seem to nod the same way.

The flowers on each stalk open strictly in sequence, with the bottom bells beginning the process and resulting in a knock-on effect until the final flower right at the tip finally opens. The bluebells remain at their most vivid for only a few days. The best time to catch the spectacle is just as the uppermost flowers have finally opened as the lower and older flowers soon begin to fade, wither and set seed.

The bluebells are pollinated by insects, primarily bumblebees and hoverflies, which have long enough probosces to access the nectar from inside the long, tubular flowers. A common springtime sight in a bluebell woodland is to see the bluebells bowing under the weight of a pollen-laden bumblebee as it – seemingly randomly – buries its head into flower after flower, pollinating different plants as it goes about the business of food collection. Bluebells also provide a vital nectar source for the pearl-bordered fritillary, one of Britain's fastest-declining butterflies, a woodland specialist that is the first of all the fritillary species to emerge and take advantage of this

Where the Spanish Armada failed, the Spanish bluebell has succeeded. The Hispanic garden escapee (left) has begun to invade Britain's woodlands, and hybridised with our native species (right) to produce a fertile cross (middle).

early nectar bonanza. It is thought that, in the event that cross pollination should not occur, bluebells are self-fertile, although the seed-set will be lower per flower spike in this case.

The bluebell owes its perennial nature to a subterranean bulb, which contains all the necessary reserves to provide the plant with a head start before it is able to photosynthesise on the emergence of its leaves. The high starch content of these small white bulbs makes them very sticky, and they were collected to make a form of glue and also used to stiffen the elaborate ruffs worn by the landed gentry in Elizabethan times. The bulb is renewed each year: as the leaves quickly die after flowering, the new shoots, flowers and roots will already be developing inside an embryonic bulb sheathed inside the current bulb, which then begins to wither away.

In addition to being able to set seed, bluebells are also capable of reproducing vegetatively by multiplication of their bulbs through a budding process. It is thought that seeds may take five years to eventually form a bulb; the addition of new bulbs to the population is countered by the fact that each year a number of the older bulbs will die off as they become unable to produce a shoot long enough to reach the soil surface.

Bluebells are not just confined to woodlands. Parts of Britain's coast and many of its offshore islands become daubed violet-blue during May.

The reason why Britain is the best place to see bluebells is simply down to its geographical position off the edge of Europe, which results in wetter and milder winters than the much colder, dry winters experienced on the Continent. British bluebells can be found anywhere from the northern tip of Scotland, with the exception of the Orkney Isles and Shetland where they are curiously absent, to the very edges of the sea cliffs in southern England. As spring generally arrives in southwest England first and is considered to proceed in a northeasterly direction at 'walking pace', it is no surprise that southern bluebells will flower well before their Scottish counterparts.

While picking bluebells is clearly undesirable, as well as being illegal, the real damage to bluebells can occur by trampling, as the damaged leaves are then unable to provide enough sustenance to ensure that the new bulb is able to flower the following year. While bluebells are covered under the 1981 Wildlife and Countryside Act, further protection was added in 1998 to prevent the wholesale stripping of bulbs for the gardening trade, an activity that is capable of decimating populations.

A more insidious recent threat is from the introduced Spanish bluebell that was brought to Britain in around 1680 and was first found growing in the wild in 1909. The Spanish species is considered very vigorous and will readily hybridise with our native bluebell to produce a fertile hybrid. This cross-breeding will undoubtedly dilute the unique characteristics of our bluebell and a recent survey by the charity Plantlife revealed that this current problem is so widespread that one in six of British woodlands is thought to have been contaminated by the Spanish bluebell or the hybrid.

While the Spanish bluebell is relatively easy to identify – it is a more upright flower, with broader leaves, more variable flower colour and blue coloured anthers – the hybrid has taken on characteristics of both parents and so can be more difficult to eliminate from the wild. In addition to the Spanish threat, climate change may present a serious future problem as the trees respond to warmer spring temperatures and open their leaves earlier; this will lead to the bluebells being shaded out during the vital growth period.

Despite these perils, bluebells are still quite rightly considered one of our national treasures and a number of walks are regularly operated at various British woods, in addition to a 'bluebell service' held in a Leicestershire wood and the famous bluebell railway in East Sussex. This is all the more reason then to celebrate and protect this wonderful and uniquely British spectacle, which announces the arrival of spring with its awe-inspiring, yet ephemeral display.

Bluebells and oak
woodland go together
like Ginger Rogers
and Fred Astaire.

Dawn chorus

WHEN
Mid-April to late May is the
best time

WHERE
Anywhere, with woodland
the most impressive

6 Dawn chorus

Bird song is surely the most beautiful sound produced in the natural world: the sheer range and variety of the birds' vocal abilities has inspired poets and musicians from Chaucer to Wordsworth and Handel to Vaughn Williams. The way to appreciate bird song at its finest in Britain is to rise before dawn and spend a couple of hours in a woodland in mid-May; it is as thrilling as it is deafening.

Although bird vocalisations can be heard in every month of the year, bird song is at its biggest and best when both resident and newly arrived migratory birds come together in spring to prepare for the cut-and-thrust of the oncoming breeding season. With a few exceptions, it is mostly the male birds that sing, behaviour that is triggered when their testes begin to secrete the combative hormone testosterone. Singing is all about communication, and is the perfect medium for transmitting information both over a range of distances and in an environment in which your rivals or potential partners may be obscured by vegetation. As sound has the ability to travel in all directions and penetrate through or around objects, it enables the males to announce one of two messages: either 'go away!' or 'come hither!'

One of the most amazing facets of bird song, in species such as the skylark or sedge warbler, is their ability to sing incredibly varied songs for long periods, seemingly without pausing to catch breath. These impressive vocal skills are produced by the birds' voice box, an organ called the syrinx, which is the avian equivalent of the human larynx. In both humans and birds, voice boxes contain membranes that vibrate and generate soundwaves when air from the lungs passes over them, with the muscles controlling the details of the sound production. However, whereas the human's larynx is situated at the top of the windpipe, the bird's syrinx is positioned much lower down at the junction of the two bronchial tubes leading to the lungs. This means that the syrinx has two potential sound sources from which to draw air, with the separate membranes on each bronchus able to produce distinct sounds, which are then mixed together higher up the vocal tract. The birds also manage to sing in sustained bursts for minutes by cleverly taking a series of mini, shallow breaths, which are synchronised with each syllable they sing.

The first reason birds sing in spring is to defend a territory from potential rivals and intruders. The retaining of this territory is crucial as it will help to determine the owner's breeding success by providing him, his mate and their chicks with their food. Males claim a piece of terrain as their own by singing in it, which serves to inform other males that it has been occupied, and to warn them from infringing the delineated borders. Many birds staking their claim in this way will frequently leave plenty of gaps in between singing to enable

The early bird (in this case the robin) gets the worm. Singing is a very serious business designed both to attract a mate and to defend a territory.

them to listen out for any replies. By constantly gauging the location of their neighbours and recognising their songs, they will immediately be alerted on hearing any strange males, the birds which will be keenest to try and muscle in on one of the currently occupied territories. These defensive songs tend to be short and simple – they are not designed to impress but are an aggressive statement that intruders will not be tolerated. Some species of bird, like the great tit, have developed a large repertoire of these simple calls; it has been suggested these are designed to convince potential rivals that the carrying capacity of males in the area has been reached, and all the territories are fully occupied.

The second reason for singing is, of course, to advertise for, and attract, a mate. When in full 'showing-off' mode, the males will sing songs that tend to be more operatic, longer and more complex than the repetitive phrases used for territorial defence. Singing is an energy-intensive activity and is designed to give the female a true assessment of how fit and strong each respective male is. So any male capable of producing a long, loud and complex song early in the morning, after a night without feeding and when his energy reserves will be low, must have an excellent territory in which to forage and must also be in exceptional condition, attributes the female will be looking for. In contrast, weaker males that are unable to summon up the energy to sing will find it difficult to hold a territory and keep a partner.

In order for singing males to further enhance the complexity of their songs for the benefit of any listening females, a number of species will imitate other species' songs in a bid to improve their repertoire. Males that have learnt the ability to mimic other species will also be able to illustrate to potential mates that they are experienced enough to have heard a wide variety of other bird species. The British master of impersonation is the marsh warbler, which has been noted copying elements of songs from over 70 bird species. This virtuoso performer even possesses the ability to mimic bird songs it has heard in its African wintering quarters, which may all help in fostering interest from a potential mate that winters in the same area. Another excellent imitator, the starling, has been known to mimic a whole variety of sounds extensively: from sheep in Shetland to curlews on the Somerset Level and car alarms in an assorted range of towns and cities!

In addition to mimicry, one way that certain species are able to increase their own repertoire is to vary the order or sequence of their 'musical phrases'. The nightingale is perhaps Britain's most famous songster, renowned for its rich and apparently endlessly variable song. This is achieved by the ability to produce more than 200 different phrases, which are then strung seamlessly together one after another in an infinite number of combinations, making each song unique. The females will then listen carefully to each performer as they visit the respective territories before choosing the classiest singer.

Natural selection has resulted in birds producing the sounds that are most effective for their respective habitat. In forest, for example, where the sound can reverberate off trees and be absorbed by the dense vegetation, the clear, pure, ringing tones of species like the blackbird are believed to carry much further distances and become less distorted than the buzzing and trilling calls of a bird like the sedge warbler. The song of the sedge warbler, however, sweeps up and down over a much wider frequency range and is believed to be better suited to a more open habitat such as that of reed bed and wet scrub in which the bird

The wren: a tiny bird with a mighty song.

breeds. Sound also travels better in the layer of air above the vegetation, which explains why so many birds sing on high, exposed perches. However, despite the message being carried much further in these loftier positions, prolonged, exposed periods leave the songster more vulnerable to attack from predators such as sparrowhawks or goshawks; some singing birds may use this to further demonstrate their brazen audacity to any females.

In order to further boost the distance their song is likely to be heard, some species such as the tree pipit and whitethroat will leap off vegetation and sing from mid-air. The most famous proponent of this aerial singing, however, is the skylark. As perches from which to sing are few and far between in the skylark's chosen habitat of open farmland, it soars into the air to deliver its glorious song. In addition to using its song both to proclaim its territory and to attract a mate, the skylark will sing even when being chased by a merlin or peregrine; this is an act of bravado designed to show its pursuer it is wasting its time because of the excellent condition of the skylark.

The dawn chorus always commences before sunrise – as the sky begins to lighten and well before the sun is seen rising from below the horizon – and is at its most intense for around two to three hours. Early in the morning the air is often still, so sounds at dawn are believed to be twenty times more effective than those produced at midday, with the best conditions for sound transmission being warm and humid mornings. Dawn is also the first time that any vacant territories may become apparent in the case of any overnight mortality. Another possible reason for this early activity is that females tend to be at their most fertile at dawn as this coincides with the period when they lay their eggs, so singing at this time may also be a form of mate guarding. The obvious disadvantage of singing during the early morning slot, however, is that the airwaves can be very crowded as lots of birds try to make themselves heard. As the morning slowly progresses the birds will then need to spend time feeding so the singing intensity falls away.

Unsurprisingly, most unmated birds sing much more often than their paired neighbours; for example, the single male pied flycatcher was found to sing around 3,600 songs a day, in contrast to the mated males close by, which delivered their song less than 1,000 times. Birds like the chaffinch and robin will also reduce their singing rate in spring on acquiring a female and only begin to sing regularly again if their mate is lost during the breeding season.

Anyone who is familiar with the structure of a dawn chorus will notice that there seems to be a systematic order of when birds begin singing in woodlands, with species like the robin, blackbird and song thrush all singing much earlier on when there may scarcely be any light at all. It is thought the birds that sing the earliest are those best able to see at low light levels as they have proportionately larger eyes. Species such as blackbirds and robins also rely heavily on earthworms in their diet, and, as the worms will be much closer to the surface in the morning, these birds have evolved to rise early to 'get the worm'; therefore, they will also be the first to sing.

Other dawn songsters such as the wren, blackcap, chaffinch and the various tit species have smaller eyes relative to their size, meaning they may not be able to spot potential predators like owl. Also, being primarily insectivorous birds, their food may not yet be active in the cooler temperatures at dawn, giving these species little incentive to rise early, which means that they are later arrivals to the dawn chorus.

Meadows

WHEN
From April to July

WHERE
North Meadow National Nature Reserve, near Cricklade, Wiltshire (Natural England); Kingcombe Meadows, near Dorchester, Dorset (Dorset Wildlife Trust); Muston Meadows National Nature Reserve, near Mustol, Leicestershire

5 Meadows in summer

The quintessential image of a meadow is waving heads of grasses interspersed with cream, purple, yellow and pink scented flowers, complemented by a vast range of humming insects. Though one of our finest summer spectacles, meadows have been decimated in the last 70 years and little room has been left for these flowers and their attendant insects in the march towards the agriculturally productive, but largely sterile, countryside of modern Britain. With the few fine intact meadows still at risk, these traditionally managed habitats have become among our most sensitive and rarest of wildlife haunts.

The nodding heads of snake's head fritillaries floating in a sea of dandelions at North Meadow in Wiltshire.

While some people think meadows are the result of leaving nature to do its own thing, hay meadows are in fact the product of long-term human intervention. As its name suggests, a hay meadow is a very traditional type of grassland that is managed in a low intensity manner – firstly for the production of hay and secondly for grazing livestock – and has not been ploughed or fertilised (other than by manure) in living memory. Unlike pasture – which represents a subtly different type of grassland that has animals present all year round – the hay meadow is free of livestock between the late spring and mid-summer. This allows the wild flowers to bloom and to set seed without being grazed or trampled. It is not until late summer, when the ground-nesting birds have reared their young and invertebrates have completed their life cycles, that the meadow will then finally be cut and the hay crop turned into bales for winter fodder. Many of the cropped plants, such as the grasses, will still continue to grow throughout the autumn and winter, allowing the re-introduction of livestock on to the grazing meadow. This period of grazing and the stocking levels will vary according to each site, the type of farming practiced and local customs, but, providing the site does not become too wet and can withstand the pressure, the livestock may stay in situ all winter.

Crucially, with this form of meadow management, the grazing actually allows the wild flowers to thrive by keeping the more aggressive species, such as grasses and shrubs, in check, and preventing the site from being overrun with scrub from the surrounding hedgerows. Additionally, the hay meadows will not have been subjected to modern fertilisers, which would overload them with nutrients, single-handedly marking their death-knell.

Some of the best and largest British sites are managed as common land, where traditional legal customs dictate strict grazing and cutting regimes. This has been the case at North Meadow in Wiltshire and has given rise to surely one of the finest remaining lowland meadows in Europe. Here, the local people of Cricklade have managed the site in a traditional manner for hundreds of years with the result that the substantial grasslands contain Britain's largest population of snake's head fritillaries, a plant that has disappeared in all but a handful of other sites. As the reserve is situated between the Rivers Thames and Churn, it frequently becomes flooded during the winter, resulting in the deposition of alluvial silt from the rivers; this acts as a natural fertiliser vital to the growth of many of the plants. The meadow dries out in early spring, and in April and May there is a showing of hundreds of thousands of the famous purple and white fritillaries, alongside yellow clusters of marsh marigold and pink patches of cuckooflower in the slightly wetter areas.

As spring changes to summer, the fritillaries have all but faded, but the meadow explodes once again with a later collection of flowers providing more blazes of colour: the purples of greater burnet and common knapweed; the yellows of meadow buttercup, cowslip and yellow rattle; the whites of ox-eye daisy and meadowsweet; and the greens of over twenty species of grass. As these flowers and grasses finally begin to turn yellow or brown and set seed, the hay is then cut any time after 1 July, and removed from the meadow before 12 August, by which time the site has often become too wet and treacherous for heavy tractors. The meadow is then turned over for use as common land for the villagers to graze their livestock, water levels permitting, until 12 February the following year. After this date the meadow is allowed to recover in preparation for the oncoming flowering season from which the next hay crop will be taken.

While the amount of grassland nationwide has not dropped since the largely pre-mechanised farming of the 1930s, the ecological value of many of the sites has plummeted. Modern farming resulted in many hay meadows being converted to much more productive grasslands through a combination of ploughing, the application of large amounts of inorganic fertiliser and the sowing of a small range of genetically improved and high-yielding species. These intensively managed species, such as perennial rye grass and white clover, prefer the churned-up and enriched soil and produce a more productive fodder for cattle than the traditional meadow, resulting in the converted meadows becoming bland, uniform and species-poor monocultures. Local wild flowers, with their precise ecological requirements and restricted distributions, disappear as they are either poisoned by the extra nutrients or out-competed by the sown species. In this way huge numbers of meadows were destroyed and, only when a nationwide survey of 'unimproved' or traditional meadows was carried out in 1984, did it become clear that just 3 per cent of the total 1930s figure was still there 54 years later. Criminally, the current total is thought to have dropped even further to around 2 per cent.

The loss of 98 per cent of our meadows can, of course, be partly attributed to their destruction at the hands of development or neglect, but by far the largest loss has come from the application of fertilisers, which has seen a whole array of once commonplace wild flowers become very localised or even rare. The few meadows that managed to escape 'agricultural improvement' often survived in nooks and crannies that could not be traversed by machines, such as steep slopes or ground that was broken up by boulders. A number of wetter meadows

on flood planes that could not be easily drained were also preserved, as were meadows with underlying clay as the soil was difficult to plough. Many of the best sites have now been protected, but even these sanctuaries for wild flowers are effectively islands in a sea of intensively farmed fields, meaning species are not be able to recolonise the meadow from the surrounding landscape.

As many meadows are often composed of plants with exacting requirements, an experienced botanist can take a look at the plant list of most sites and guess the geographical location of the meadow and its underlying geology. Plants such as the snake's head fritillary and autumn crocus, for example, are classic meadow species in the south of Britain, while flowers such as globeflower and wood cranesbill are mostly confined to northern unimproved grasslands. Other rare plants virtually confined to hay meadows include the green-winged orchid and adder's-tongue fern. In contrast to the species-poor communities that comprise today's modern grasslands, the hay meadows are far more diverse, with 30 species per square metre at the richest sites, and many meadows comfortably containing at least 120 species. All these plants are able to co-exist as they grow at different times of the year and, therefore, have different requirements for light, water, nutrients, and specific pollinating insects.

Meadows also have an incredibly diverse fauna, including rare butterflies, such as the marsh fritillary and marbled white, a plethora of insects, and insects that eat insects. These, in turn, will attract small mammals, such as voles, shrews, hedgehogs and bats. Meadows also make good breeding sites for birds like the skylark, yellow wagtail and grey partridge, and, in the winter, sites that flood can be valuable for waders, such as lapwing, snipe and curlew.

In addition to being important wildlife refuges, meadows are an iconic part of our rural heritage and a highly valued feature of the landscape. So, even if you are not predisposed to get to grips with the finer points of grass identification, why not value these special places as a source of tranquillity, inspiration and a break from the hustle and bustle of modern life?

Black grouse

WHEN
Early mornings in March and
April

WHERE
Ruabon moor, Llandegla
Forest Centre, near Wrexham;
Galloway Forest Park,
Dumfries & Galloway
(Forestry Commission);
Abernethy Nature Reserve
(RSPB), Inverness

4 Lekking black grouse

It is surprising the number of keen birdwatchers who have never made the time to visit a black grouse lek; surprising, in that it is surely one of the most fascinating behavioural spectacles this island has to hold. Unfortunately, seeing the males or blackcocks strutting around their lek at first light in early spring has now become a rare sight, as this species has suffered a steep and precipitous decline over the last century.

The black grouse survives as a bird of uplands, most often on moorland and hill farm habitats, where a mosaic of rough pasture, bog and young conifer plantations are to be found. Now confined only to the Welsh uplands, the north Pennines and remote parts of Scotland, it is hard to believe that a century ago, black grouse leks could be found within a few dozen miles of London and were commonly seen from Norfolk through to Hampshire and Dorset to Cornwall.

Even in their strongholds, the black grouse is a surprisingly inconspicuous bird for most of the year; the glaring exception is during a few short weeks between mid-April and early May when the males gather at their ancestral arena or lek. Early birdwatchers thought the fights on these leks were pointless, but they have been discovered to be anything but, with the victors winning the ultimate animal race: the right to pass on their genes to the next generation.

Leks are usually located in an open area at the centre of a peat bog, moor or rough wet grassland. It is an area that is easily accessible to the surrounding birds and must also be relatively undisturbed. As the location requirements are quite exacting, it is no surprise that a number of leks have been used and reused for over 50 years. The older males have a particularly strong attachment to sites and this in turn attracts the younger birds, which maintain the tradition.

The size of the lek varies according to the number of displaying males; anything above a dozen regular males is considered to be a large lek by contemporary British standards. This is minuscule compared to some of the leks in the Soviet Union, which have as many as 200 individual birds.

At first sight it can be confusing trying to interpret the seemingly complex behaviour at a lek, as the birds noisily jump around and display to all and sundry. However, with some persistence, it soon becomes obvious that there are distinct patterns to their behaviour. Each male that regularly visits the lek has already carved out a clearly delineated territory. The dimensions of these territories hardly vary from day to day and can easily be mapped out by watching where the males fight each other along their respective borders. The larger the number of males at the lek, the smaller the respective sizes of each territory. In the centre, the territories are the smallest and are occupied by the dominant males who have the most success when it comes to attracting the females. The subservient or less-experienced birds are always pushed to the

Dressed to impress. Two neighbouring male black grouse face off across their shared territorial boundary.

Key to the black
grouse's survival is
to be left undisturbed
at their ancestral
leks. These breeding
grounds are situated
on peat bog, moor or
rough wet grassland.

periphery of the lek, where they often hold much larger, marginal territories with no external boundaries.

Besides the birds that regularly attend the lek, a small number of males called 'incidentals' occasionally watch the proceedings. It is thought that these birds are immature males that have been unsuccessful in proclaiming a territory and so turn up to learn from the more experienced performers. There is additionally a third type of male that never attends a lek at all; these birds – 'soloists' – seem to prefer performing on the ground or in a tree alone in the hope of trying to entice the occasional female away from the clan.

The males usually appear at the lek site before dawn and, often, the first time the human observer will be aware of a bird's presence is from the whir of wings as it arrives or a strange two-note crowing or hissing call it emits on reaching its territory. This far-carrying call is best described as 'choo-iischt' and is used by each male to announce his presence and position on the lek. While making this call, the male also fans and raises his lyre-shaped tail, exposing the white rosette-shaped cluster of undertail feathers, and extending and holding his head and neck upright. This crowing noise is often made while performing little jumps on flapping wings, so that the flashing white undersurface of the wings adds to the effect. This call is thought to be used primarily to advertise the location of the lek to females within earshot.

Soon after arriving at his territory, the male will then often thrust his head and neck forwards, and, with his crimson eyebrows raised and neck swelled, utter a far-carrying bubbling dovelike 'rookooing' song. This noise is produced by air rushing from the bird's chest to an inflatable oesophagus, and, although each call only lasts just over two seconds, the calls often roll together into a long set of bubbling notes. Each of these individual calls is produced while the male's beak is closed and he only opens his beak briefly after each burst to draw in a new breath. When the male starts 'rookooing', the effort causes his whole body to shake, and he will also drop his wings to reveal a series of white studs on the leading edge. This is to make himself seem even more impressive. The 'rookooing' song is primarily considered to be an aggressive display, as males often produce the noise alternately at one another across a territorial boundary in 'sabre-rattling' fashion. This song is surprisingly powerful and can carry for distances of over 400 metres. It also has a ventriloquial quality, making it difficult to pinpoint the source of the sound. In addition, it may also stimulate the female into preparing for breeding and egg production.

Both the crowing and 'rookooing' are used extensively in defence against neighbouring males and interlopers. When scuffles occur along the territorial boundaries they are often only display fights with both males displaying and then one taking a quick run at the other while the other retreats and vice versa. Most encounters end without any contact, but birds will occasionally strike each other with their beaks or wings, and have even knocked over and trampled on a neighbour in fights that get out of hand. Males intruding well inside another's territory are often driven off mercilessly by the resident male.

When females arrive at the lek, they will often sit for long periods in a nearby tree or on the ground and nervously watch the proceedings unfold. When finally a female is ready to participate, she walks around the lek, causing the males to go into a frenzy of displaying and posturing. Once the female moves into a particular territory, the resident male begins 'rookooing' constantly, while encircling her with extremely rapid steps and tilting his tail towards her as he passes in front. This behaviour is often called 'courting', but it is thought to be more of a threat to demonstrate his superiority to the female and to bring about her complete subordination. The dominant males in the centre of the lek have honed their technique over a number of breeding seasons and tend to have much more success than the envious bystanders.

Despite the male's dominant display, it is the female that dictates when she is ready to mate: after having been circled by her chosen male a number of times, she crouches with a squat, submissive posture and a flat tail. The male takes this as his cue, mounts her and mates her with an aggressive exuberance, flapping his wings for balance and brutally seizing the nape of her neck.

The venerable ornithologist David Lack showed how hyped up the males are during the height of the breeding season. He placed a stuffed female or hen bird in the middle of a lek and was then able to watch one testosterone-charged male mate with the stuffed bird a total of 56 times in a 45-minute period!

To see a lek in its full glory requires some dedication. It is important to arrive before the birds, so a cold, early start is essential. The grouse are very susceptible to disturbance so a reputable guided walk operated by wardens who know the birds intimately is highly recommended. We have already decimated their habitat and shot them in huge numbers, and it would be a shame to see the remaining 6,500 males (the females are so secretive their numbers are unknown) suffer any more purely because of our overexuberance to see one of Britain's most impressive mating games.

3 Leaping salmon

While fishing is quite comfortably our most popular pastime, the art of fish-watching figures low down the pecking order for the average naturalist, compared to that of, say, birdwatching. However, for a couple of weeks in the autumn, a wondrous fishy display takes place, which temporarily relegates anything our feathered friends can muster to no more than a supporting act. Seeing a salmon run in full flow is simply one of the most exhilarating spectacles that you will ever hope to witness.

The return of salmon from the sea to breed in the exact river in which they hatched is one of the greatest, yet mostly unheralded, migration stories; it is also a phenomenon that marks the end point of the lives of the vast majority of salmon. The reason this spectacle usually plays second fiddle to the arrival of the first swallow or the appearance of our winter geese is down to the fact that many of the secrets of their complex life cycle have only recently been unravelled.

The salmon that negotiate their way to the sea unscathed will leave the freshwater behind for the productive yet dangerous waters of the North Sea, where they will spend between 15 and 50 months increasing in size and maturity in readiness for the return leg of their journey. The young salmon – or smolt as they are called at this stage – will spread out across the North Sea, travelling anywhere from the Faroe Islands to the Norwegian Sea and southwest Greenland as they feed in the rich deep-sea waters on a diet of crustaceans, post-larval sand eels, and small fish such as capelin and herring.

Having spent the winter at sea, some smolts return to breed the following autumn, but salmon from Scottish rivers will often spend at least two to three years putting on weight out at sea before the urge to breed drives them back to the freshwater. While the food at sea is plentiful for the maturing salmon, there are many predators – sharks, seals, dolphins, killer whales and man – and only ten per cent of the smolts that initially leave their rivers will ever return.

At the onset of maturity, the fish begin to make their journey home; it is one of the great marvels of nature. The way that the salmon can pilot their way back to their native rivers has not yet been completely solved, but it is thought to involve guidance by stars, use of the Earth's magnetic field, and a knowledge of the ocean's currents. Once back at the coast, they probably locate their natal river by using a 'chemical memory' of a variety of substances in the water such as pheromones, although these are only present in tiny concentrations.

When they finally arrive back at their estuaries in preparation for the big push up the rivers, the sexes of the silvery coloured smolts are virtually indistinguishable, but, after a few weeks, they begin to develop their breeding 'tartan', as it is called in Scottish waters. The females become much darker all over and have a pronounced rainbow colouration along the flanks, while the

Once the salmon enters freshwater it is a lean, mean swimming machine, determined to risk life and fin to reach the spawning grounds upstream.

Salmon

WHEN
In October and November,
and best after heavy rains

WHERE
Buchanty Spout near Crieff,
Perthshire; Linn of Tummel,
near Pitlochry, Perthshire
(National Trust for Scotland);
Ettrick Weir, River Tweed,
Berwick upon Tweed, Borders

males develop red bellies and a streaking pattern along the flanks, which is as individual to each fish as a tartan is to each Scottish clan. Additionally, the head of the male salmon becomes elongated and it develops a large hook called a kype, which protrudes upwards from its lower jaw and will be used against other males to ensure mating rights further upstream. In Scotland, once the salmon come into their breeding dress, the adult males and females become known as cocks and hens respectively.

For the duration of the adults' time in the river system they will rely solely on reserves put down during their time at sea. In freshwater there are few predators (apart from man) capable of catching and eating such large fish: many adult salmon by this stage comfortably surpass 60 centimetres in length and 7 kilograms in weight. The largest ever caught by rod and line was a 29-kilogram monster that was reeled in by Georgina Ballantyne on the River Tay in 1922.

Once the adult salmon enter the river system they travel against the current for the entire journey and have to use their strength and agility to power up the river. Movement of the salmon is often dictated by rainfall and the resulting river flow; the river level plays a critical role in enabling the toiling salmon to work their way past a variety of obstacles as they proceed up the river.

Of course, the spectacle for which salmon are justifiably famous is when they leap up waterfalls and momentarily break free of the water as they are mercilessly driven on by their Darwinian prerogative. The height that salmon can leap depends upon the relative depth of the water at the foot of the waterfall and the creation of a 'hydraulic jump' that is caused by the initial downwards force of the cascading water, which in turn creates an upsurge to give the salmon an initial push. By this stage, the fish are rippling muscle, and they use this to drive up from the river bottom in a short sprint, which helps them to power up through the water and into the air. If they manage to leap clear to the top of the watery obstruction and land back in the water they will then propel themselves quickly up and through the rapids by vigorous contractions of the tail fin. Many leaps seem to end in heroic failure with the fish often getting thrown back into the churning bottom of the waterfall, but the vast majority will eventually successfully pass the obstruction, particularly after heavy rainfall, when the flow is higher. These bottlenecks also make the best places to watch the salmon, with early mornings and late afternoons being when the fish seem to be most active.

Once all the obstacles are negotiated and the surviving salmon arrive at the stretch of river that they were born in, they are keen to get about the business of mating. Spawning dates can vary between rivers and are influenced by water

Once the obstacles have been negotiated, the salmon are able to begin spawning at sites like the Upper Tweed in the Scottish Borders.

temperature and the amount of daylight. Generally, however, it takes place between November and December in Britain. Firstly, the hen salmon digs a series of shallow depressions in the gravelly bottom of the river; these are called redds. After she has deposited the eggs in the redds, they are immediately fertilised by an accompanying cock male and often additionally by much smaller but precocious salmon called parr that have not yet made the initial voyage to sea. The fertilised eggs are covered over with gravel by the female, who will then lay more batches, culminating in a final figure of around 7,000 eggs during the entire process. Spawning is so exhausting that over 90 per cent of the adult salmon die at this stage, although a few post-spawning salmon, or kelts, do seem to return to the sea to spawn again in future years.

Hatching of the pale orange eggs usually occurs early in the following spring and the young fish, or alevins, remain initially in the redd for a few weeks as they feed on their attached yolk sac. When they do emerge from the gravel as fry in April or May they are only two or three centimetres long. The fry then begin to feed on aquatic invertebrates, but there are often far more juveniles than the river can sustain meaning mortality can be as high as 90 per cent. Once the surviving fry have grown and become well patterned along their flanks, they develop into parr and spend the summer feeding in the fast shallows and riffles as they continue to grow. Goosander, pike and herons, in addition to the danger of the river drying up in a summer drought, all take a further toll.

Normally the parr will feed for at least a year – more in Scottish rivers – before they are ready to go out to sea for the first time. Their departure is marked by losing their distinctive markings as they turn a silver colour and become smolts. Each spring the smolts will start to drift downriver at night in shoals and are once more vulnerable to predators, particularly at estuaries where large numbers are picked off by cormorants and gulls.

This entire journey is so arduous that only one in a thousand fertilised eggs will last all the way through to become a spawning male, but, even in bad years, there are usually enough salmon to restock the rivers. While stocks of salmon have been in general decline in recent years because of pollution, obstacles to migration, fisheries and climatic change, with the salmon still leaping there is much hope for this most resilient of fish species.

While many leaping salmon often seem to be thrown back into the tumbling waters below, the vast majority do eventually make it upstream to spawn.

Starlings

WHEN
Best from November to
February, although some
roosts can be ephemeral

WHERE
Westhay Moor Nature
Reserve (Somerset Wildlife
Trust); Slimbridge (WWT);
Brighton Pier

2 Roosting starlings

The venerable zoologist VC Wynne-Edwards said of the starlings' aerobatic display: 'it seems quite irrational to dismiss what is certainly the starling's most striking social accomplishment merely as a recreation devoid of purpose or survival value, and wiser to assume that a communal exercise so highly perfected is fulfilling an important function.' Indeed, watching a swarm or 'murmuration' of starlings swirl like a huge animated cloud of smoke before entering their roost is simply one of the greatest ornithological highlights Britain offers. And while, as Wynne-Edwards also agreed, the spectacle can be enjoyed purely at the level of a visual feast, it is a far more satisfying experience if you understand more of the behaviour underpinning the display.

It is most definitely a case of safety in numbers when the local peregrine appears at a starling roost intent on catching dinner.

Whereas the huge starling roosts that are the feature of this spectacle are essentially an autumnal and winter phenomenon, it is also worth bearing in mind that starlings do roost communally throughout the year. During the breeding season, however, roosts are small and contain mostly non-breeders and some breeding males. To see the biggest and best roosts requires a combination of the correct timing and the usual portion of 'wildlife luck'.

The actual size of the largest of these autumnal flocks can be truly mind-boggling. While impossible to count because of the number and movement, it is estimated that the number of starlings coming to roost at Britain's greatest starling roost location, Westhay Moor in Somerset, is approximately 7 million birds. Pick a good night with amenable weather conditions and, at a number of known locations throughout Britain, you are in for a treat as the birds put on a virtuoso display of mass, synchronised flying that can blot out the sky and leave your jaw on the floor.

Despite a recent downturn in numbers, the starling is still comfortably one of our most common breeding birds, with most recent estimates putting the breeding population at around 1.1 million pairs. Therefore, as the entire British breeding population could fit three times over into the Westhay roost with room to spare, it is obvious that our resident birds are supplemented by a large immigrant population in October and November. Attracted to Britain's mild Gulf Stream climate and our position on the edge of the Continent, huge numbers of Continental starlings cross the North Sea from the Netherlands, Germany, Scandinavia and even as far away as Russia for a winter holiday as they blend in with our resident population.

The largest roosts in Britain are now mostly confined to rural areas in habitats such as small woods or reed beds. Historically, Trafalgar Square had a famous starling roost, but the noise they caused and the mess, which often covered buildings of considerable architectural heritage, meant that the reactions of many Londoners were decidedly mixed. The population of London's starlings in the middle of last century was so large that, on 12 August 1947, Big Ben was actually stopped by the weight of birds sitting on the minute hand! Unfortunately, the huge gatherings of starlings in cities such as London, Glasgow, Liverpool and Bristol now seem to be a thing of the past.

During a typical autumn or winter's day, the starlings spend their time feeding away from the roost at distances of up to fifty miles, although many birds are thought to feed no further than twenty miles away. As the afternoon proceeds, the starlings then begin to gather in 'pre-roost assemblies', usually within three miles of the main roost site. The locations of these assemblies are often at traditional sites that represent good feeding areas, and, during this pre-roost time, the starlings not only feed intensively to ensure they have enough energy to survive the night, but also use this time to sing, bathe and preen in preparation for a good night's sleep.

These pre-roost assembly sites are often locations such as pig or cattle farms, where food is very plentiful and pickings are easy as the starlings help themselves to food laid out for the stock. As early evening commences the starlings will leave these 'pit stop, refuelling stations' for the main roost and, as they pass over other small flocks stocking up at locations closer to the roost site, these flying flocks stimulate the feeding birds on the ground to fly up and join them. This pattern is repeated for miles around and, with dusk rapidly approaching, there is definitely a feel of 'the gathering storm' as the flocks coalesce with other groups while converging on the roost, like the spokes on a wheel.

Displays over the roost can sometimes be brief or even occasionally non-existent in very wet weather, but, if conditions are favourable, the first small flock of starlings to arrive at the roost will often form a small, tight flock and weave around above the roost site as the sky begins to darken. This display acts as a 'flag' or focal point advertising the location of the roost for other flocks coming in and also as an invitation to join the flock. As the arriving flocks fly in from all points, the initial flag, consisting of as few as 50 birds, swells quickly.

As larger and larger flocks join the flag, the swarm above the roost takes on a life of its own. From a distance the flock can look like smoke and is ever-changing in shape as the tightly knit swarm grows and changes direction rapidly. From below, the swarms can become so expansive that they do indeed seem to blot out the sky and, as they fly overhead in perfect formation, they often seem to resemble an MC Escher image!

Close at hand, individual birds can be picked out quite literally wing-tip to wing-tip as they constantly alter their flight paths at a second's notice with effortless ease and astonishing technical ability.

As the 'mother flock' continues to double and treble in size, satellite flocks carry on streaming in from the surrounding countryside and, at this stage, the only sound that can usually be heard from the birds above is the swoosh of the wind as thousands of pairs of wings change direction in unison. When the flock is so large it becomes very difficult to count as it is constantly morphs into different shapes, becomes multi-layered and whips around in seemingly random directions at speed. I once tried to estimate the population at a large reed-bed roost and ended up discussing my frustration about the nigh-on-impossible nature of my job with a fellow birdwatcher, who told me, 'It's easy. Just count the number of wings and divide by two!'

Of course, such a huge population of starlings in one place is sure to attract predators and at large roosts it is not unusual to see at least three or four sparrowhawks and maybe even a peregrine attempt to take advantage of what appears to be a superabundance of food. However, this is where being in large numbers becomes a considerable advantage, as a starling flock of, say, 100,000 with 4 sparrowhawks in attendance, will reduce their potential chance of being caught to as low as 25,000 to 1 as the local predator population becomes swamped. In addition, sparrowhawks, for example, often need to focus on one particular bird to chase down and they find it difficult to isolate an individual from among the confusing, swirling mass in front of them, further lengthening the starling's odds of capture.

As late arrivals continue to stream in and dusk rapidly approaches, the main

Starlings have deserted our cities, so one of the very few urban roosts left to admire is at the iconic West Pier in Brighton.

flock usually passes to and fro over the roost site, with birds scything away from the bottom of the flock and aerially checking out the roost before rejoining the main flock. Suddenly the behaviour changes and it feels as if a plug has been pulled out of the reed bed as birds pour into the roost in huge swathes. Quite often, large flocks of a couple of thousand at a time may well enter the roost, cascading out of the air before the main body of the flock lifts up again, only to drop closer to the roost as another huge portion of the main flock peels away and tumbles out of the sky to terra firma. The starlings do often drop very quickly into the roost, as, having left the safe confines of the main flock, this can be when they are vulnerable to predators. As staggering numbers continue to pour out of the sky, the flying flock dwindles quickly, to the extent that, often in the space of less than ten minutes, the sky can go from being virtually obscured to completely empty.

The birds are relatively quiet during the aerial display, but as they settle into the roost the noise can reach a crescendo. There is considerable jostling for position within the roost itself and the distribution of the birds is not random. Adult male starlings tend to occupy the favoured central positions and younger birds – particularly females – tend to be pushed to the periphery, which is colder and more susceptible to attack from predators. The most dominant birds also favour the more elevated positions in the centre, as birds lower down are prone to being covered with droppings from the birds above, which reduces the waterproofing quality of their feathers.

In roosting habitats like reed beds, the starlings are very densely packed; they may only roost 15 or 20 centimetres apart (or even closer on very cold nights). The reason for roosting cheek by jowl is a combination of the substantial warmth generated by many compact bodies and the protection granted by safety in numbers. However, it has been estimated that up to 650 birds weighing a total of around 46 kilograms may be present in each cubic metre of reed bed. Clearly the flimsy reed stems are unable to support this weight indefinitely and the sheer numbers of birds do destroy certain areas of reed bed, meaning that the roost has to move location at various times through the winter. Roosts in small woods can also cause damage from the sheer amount of droppings left plastered to the trees.

In recent years this truly remarkable and under-appreciated bird has gone from being one of our most abundant breeding birds to a species that is in steep decline. It is thought that changes in land use, such as the loss of grassland to cereal production, have played an important part in the reduction of our resident population. The spectre of climate change may well also, in the future, affect the size of our winter roosts, as warmer temperatures on the Continent mean that Dutch and German starlings prefer to winter nearer to their breeding grounds rather than spend the winter flattening the Somerset reed beds.

Like all the spectacles, it is wonderful that they are there to be enjoyed but, while watching this wonderful phenomenon, remember that it is down to us to ensure they are conserved, protected and never taken for granted.

Starlings enjoying a sociable chat before heading off to the roost.

Gannets

WHEN
Any time from March to
October, but May to mid-July
is the best time

WHERE
Bass Rock in the Firth of
Forth; Bempton Cliffs Reserve
(RSPB) in Yorkshire;
Grassholm RSPB Reserve off
the Pembrokeshire coast

1 Diving gannets

With their hugely convoluted and windswept coastlines, our islands are a very special place for seabirds, and few would argue that the most majestic of all these species is the gannet. In fact the 24 remote breeding British and Irish outposts, and the rich seas that swirl around them, are so important to this acrobatic fisherman that they house an astonishing 63 per cent of the entire world population.

The gannet's long tapered body, its 180-centimetre wingspan, its snow-white plumage with black wing-tips, and its yellow head and nape make this a species that cannot be confused with any other. It is also the seabird that spends the most time intricately tied to its breeding colony: November to early January are the only times during the year that locations such as Bass Rock, St Kilda and Grassholm become gannet-free zones.

Away from the breeding sites the adult gannets spend the short winter months leading a nomadic lifestyle as they disperse into the surrounding continental-shelf waters of the North Atlantic. Despite the sometimes inclement weather, the gannets have no need to expend energy in flying to and from the colony, feeding young or mating and displaying, which means they are able to spend the winter putting down considerable layers of fat that will be needed during the rigours of the oncoming breeding season.

Hardly have the New Year hangovers faded than the gannets begin to reoccupy their ancestral breeding sites. Gannets are by nature a monogamous, long-lived species with a breeding success that is enhanced by mate fidelity; they are also creatures of habit and established pairs invariably use exactly the same nesting location each season. In a crowded gannet colony, the male that arrives back as early as possible is able to secure the pair's ancestral spot, possession being nine-tenths of the law in the gannet world. Once a patch of ground has been established by the male he will, with the help of his mate, stoutly defend it against any pushy neighbours or young pretenders with the intention of muscling in.

Unlike guillemots, razorbills and kittiwakes, which always confine themselves to ledges, gannets will nest in abundance on slopes or the flat tops of the islands. However, it is thought that the ledges are still the most desired residences: first because they give trouble-free access to the sea, and second because they make easier landing pads as the cliff updraughts make it much easier for a bird of this size – they can weigh as much as three kilograms – to land on a sixpence.

The individual nesting sites of long-standing pairs frequently develop into mounds that can reach over 50 centimetres in height as the pair continually add to their ancestral pile each season with vegetation gathered from the

It's cheek-by-jowl living in the high-rise world of the breeding gannet.

island, or seaweed and human rubbish that have been collected out at sea. An eclectic range of objects has been recorded adorning some nests, which has included, in addition to netting and fishing lines, a set of false teeth, a gold watch and a plastic frog! The nests are often so tightly packed together that the average density is rarely less than two per square metre, which means that each nest is only marginally beyond the length of their neighbour's pecking beak. When seen from a distance, the crowded colonies are so jammed together that they can give the impression of a spectacular 'snowfield' as every centimetre becomes occupied.

The pair defend their all-important nesting site by a combination of ritualised displays, and, if necessary, by a good old-fashioned tug-of-war with their daggerlike bills. The male will frequently 'threat gape' at neighbouring males, by menacingly extending and twisting his head and partly open bill in their direction. Both sexes will also regularly perform a beautiful sight-ownership display, which involves calling in unison with their wings held out in heraldic fashion as they bow and shake their heads, before ending the show by tucking their bill-tip down among their breast feathers.

Looking like a freshly baked cake just dusted with icing sugar, Bass Rock in the Firth of Forth gleams with the accumulated guano of close to 40,000 pairs of gannets.

Because it takes so long to acquire the social skills to actually gain a nesting site and attract a mate, many of the gannets will not breed successfully until they are between six and ten years of age. In addition to the mature breeding pairs, therefore, the nesting colony will invariably contain a large proportion of gatherings of birds that seem to have developed the full adult plumage but are still essentially immature. These non-breeding birds tend to gather in transitory groups called 'clubs' along the edges of the breeding colony where they practise the range of behavioural skills they'll need once they are able to carve out a nesting site.

The gannet's single white egg is laid at any time between late March and early July with the peak laying date being around the middle of April. Both sexes take it in turns to incubate the egg for a total of around 43 days. Their cargo is initially kept warm in between the webs of their feet until it is close to hatching, when the egg is then transferred to the top of the feet to ensure that the chick is not crushed inside during emergence. At this stage, the precious egg – and, when it emerges, the chick – has to be guarded around the clock as any number of potential avian predators, such as herring gulls and lesser black-backed gulls, are constantly on the alert for an easy meal. During the early period this means there has to be a constant presence of at least one parent and changeovers between the pair at the nest have to be carefully coordinated to ensure the chick isn't left vulnerable. For this they use sky-pointing. This involves pointing the bill directly upwards and is a direct communication between the pair, which can be translated as 'I'm about to leave the nest'. Once completed, this leaves the 'sky-pointer' free to go fishing.

The black, naked chick is completely helpless at birth but is able to take predigested, liquid food from its parents almost instantly and is brooded for the first three weeks until it develops a white downy covering that makes it resemble a big white cotton ball. Once the chick is able to thermoregulate, it will grow quickly on a diet of mackerel or herring that the parents collect several times a day, and, upon fledging, the youngster will have changed colour yet again as the dark flight feathers replace the light down.

A big white cross in the sky, with ink-black wing tips and a yellow nape ... this is a bird that can't be confused with anything else!

At around 90 days old, the chick is forced to fledge in order to avoid starvation as the adults stop bringing in food in an attempt to encourage their offspring to make the leap of faith. Egged on by hunger, the juvenile, still-flightless bird has to jump from the precipitous cliffs into the water before paddling south on the start of a long, lonely journey to West and North Africa where it will spend the next four years before returning to its natal colony. Initially, it is still not able to fly; a few days' swimming reduces its weight sufficiently to allow it finally to take off. The ability to catch fish is a difficult technique to acquire, and it is thought that the calmer and warmer tropical seas where the fish are generally smaller is a good place to learn the necessary skills before going fishing in the north Atlantic with the big boys.

The gannet's incredible ability to locate mackerel and herring shoals has been known for centuries. Britain's gannet expert, Bryan Nelson, in his book *The Atlantic Gannet*, quoted a Clyde fisherman called Donald McIntosh, who said 'when ye see them hingin' yon wey, cocking their nebs, that's when the herrin' wir right thick.' The gannet's huge wingspan enables it easily to travel as far as 200 miles away from the colony as they cover huge areas in search of shoals of fast-moving fish. Binocular vision and polarising eyesight also help them spot their prey even in choppy waters. When a shoal is finally located, the gannet flies into the wind to give it the necessary control as it plans its attack strategy.

Gannets dive from heights between just above the surface to an incredible 45 metres above the water, with most dives averaging around the 10 to 15 metre mark. The plunge itself is mainly powered by gravity, but, as they dive, they are capable of making minute adjustments to alter their angle and direction, sometimes even by flipping beyond the vertical where necessary to ensure a successful catch. Despite hitting the water at a tremendous speed of around 60 miles per hour, the entry is streamlined and, as the gannet's wings lie backwards along the tail, they form a perfect arrowhead as the closed bill cleaves the water. The depths reached during the dive are not usually more than 10 or 15 metres, but, once below, they can use their wings and feet to propel them the final short distance to their target. Most dives do not last much longer than five to ten seconds, in which time they will grab or spear the slippery customer with their serrated bill before their natural buoyancy allows them to return them to the surface. Most fish are swallowed underwater, unless a particularly large item has been snared; this is then polished off at the surface.

ABOVE: Despite being perfectly streamlined for entry, gannets still make a tremendous thump as they hit the water.

FACING PAGE: It is a spectacular flurry of beaks and wings...if you are a fish beware!

When a shoal is discovered, there will be more than enough to go around and the very sight of one dazzlingly white bird rising up before plunging in is enough to attract all the other keen-eyed gannets from miles around who are keen to join in the smorgasbord their colleague has located. As the bird's find quickly spreads along the gannet grapevine, a huge feeding flock can develop: hundreds of birds will plummet beak-long into the water in a series of criss-crossed dives – a truly stunning spectacle. Often a total frenzy seems to grip the gannets and there are even accounts of gannets diving at fish on the decks of boats!

Gannets are easier to count than virtually every other seabird because they nest above ground, and the future looks optimistic for our British gannets as the colonies continue to steadily increase at between one and two per cent each year.

Index

Page numbers in *italics* denote photographs.

Tiree 49
toads, natterjack *14*, 15-17
toadstools 125
trapping moths 30, *31*, 32-5, *32*, *33*, *34*
trees, autumnal *122*, 123-5
twite 51

Valley of the Rocks, Lynton, Devon 53

waders 42-7, *42*, *44*, *45*, *46*, 51
Wales 18, 37, 52, 53, 67, 80, 81, *108-9*, 142, 143, 201 *see also under individual area*
Wash, The 42

West Woods, nr Marlborough, Wiltshire 184
Westhay Moor Nature Reserve, Somerset 208
Westonbirt Arboretum, Wiltshire 122
whales, minke 135-7
wheeling puffins *116*, 117-21, *119*, *120-1*
white fritillaries 198
White, Gilbert 57, 61, 166
white-fronted geese, Greenland 50
wild goats rutting 52, *53*, 54-5, *55*
Wildlife Countryside Act 1981 188
Wilson, EO 18

Wiltshire 76, 84, 122, 184, 196, *196*, 197
winter high-tide roost *42*, 43-7, *44*, *45*, *46*
wood ants 18, *19*, 20-1, *20*
woodland 133, 182, *183*, *184-5*, *189*
woodpeckers 101
wren *193*, 195
Wynne-Edwards, VC 209

Yorkshire 122, 169, 214

Picture credits

The following images are from NHPA: p.14, p. 24/5 David Woodfall; p.19, p.22, p.32, p.59, p.74, p.76, p.77, p.78 Stephen Dalton; p.20, p.33, p.37, p.56, p.62, p.64, p.66, p.67, p.71, p.72/3, p.83, p.93, p.95, p.97, p.112, p.122, p.142, p.145, p.153, p.189, p.202, p.206, p.213 Laurie Campbell; p.24/5; p.26, p.40, p.156, p.158, p.160, p.167, p.168/9, p.170, p.200, p.205 Andy Rouse; p.29, p.138, p.146, p.149, p.151 Manfred Danegger; p.31, p.34, p.80 Robert Thompson; p.38/9, p.45 (right), p.46, p.60, p.218 Roger Tidman; p.42, p.45 (left), p.91, p.164 Mike Lane; p.44, p.193 Simon Booth; p.53, p.103, p.107 (left), p.110, p.116, p.134, p.192, p.194 Alan Williams; p.55, p.58, p.119, p.216, p.217 Bill Coster; p.68, p.108/9 Mike Lane; p.70 Eero Murtomaki; p.82, p.84 Alan Barnes; p.85 Nigel J Dennis; p.86 Jean-Louis Le Moigne; p.88 Ann & Steve Toon; p.89, p.128/9, p.130, p.131, p.132, p.148, p.184/5, p.190, p.196 Ernie Janes; p.99, p.100, p.107 (right) Iain Green; p.105, p.162, p.179 Joe Blossom; p.120, p.154 Danny Green; p.124 (top), p.124 (bottom) Yves Lanceau; p.127 (left), p.127 (right), p.199 Guy Edwardes; p.136 Trevor McDonald; p.140 (top), p.140 (bottom) Hellio & Van Ingen; p.173 John Hayward; p.174/5 Linda Pitkin; p.183 Jane Gifford; p.186 Mark Bowler; p.187 Michael Leach; p.207 Bryan & Cherry Alexander; p.208 Andrea Bonetti; p.211 James Warwick; p.214 Alberto Nardi

The following images are from FLPA: p.49 Bob Gibbons; p.50 Derek Middleton; p.61 David Hosking; p.180/1 Roger Tidman; p.219 Paul Hobson